The Institution of Mechanical Engineers is pleased to
present this volume on the occasion of its
150th Anniversary

Pam Liversidge

Pam Liversidge
President, IMechE

Visions of Tomorrow

Conference Organizing Committee

P Liversidge — Symposium Chairman; Managing Director, Quest Investments Limited; President, IMechE

Professor W Banks — Department of Mechanical Engineering, University of Strathclyde

J Cartmell — Engineering Manager, Adtranz (Total Rail Systems) Limited

I Grant — Managing Director, Volvo Aero Turbines (UK) Limited

R Harrop — Group Chief Executive, Servomex plc; Vice-President, IMechE

W Hillier — Visiting Industrial Fellow, University of Cambridge

Dr M Lord — Senior Lecturer, King's College, School of Medicine and Dentistry, King's College Hospital

Professor P McKeown *OBE, FEng* — Emeritus Professor, Cranfield University; Consultant in Precision Engineering

Professor D Sheldon — Director of Manufacturing, T&N Technology Limited

C Simpson — Divisional Chief Executive, Yule Catto plc

Professor C Taylor *FEng* — Dean of Faculty of Engineering and Director of the Institute of Tribology, University of Leeds

Honorary Advisory Panel

P Liversidge — Symposium Chairman; Managing Director, Quest Investments Limited; President, IMechE

Professor E Shannon *FEng* — Immediate Past President, IMechE

Professor D Dowson *CBE, FRS, FEng* — Emeritus Research Professor, Department of Mechanical Engineering, University of Leeds; Past President, IMechE

Professor R Brook *OBE* — Chief Executive, EPSRC

R Malpas *CBE, FEng* — Chairman, Cookson Group plc, British Co-Chairman of Eurotunnel plc

Sir Christopher Lewinton *FEng* — Chairman & Chief Executive, TI Group plc

Sir Ralph Robins, *FEng* — Chairman, Rolls-Royce plc

Sir John Fairclough *FEng* — Chairman, Rothschild Ventures Limited

IMechE
Conference Transactions

I MECH E

150th Anniversary
1847 - 1997

150th Anniversary Symposium

Visions of Tomorrow
– Improving the Quality of Life through Technology

7–8 July 1997

Organized by the Institution of Mechanical Engineers (IMechE)

Supported by:

IMechE Conference Transaction 1997–3

Published by Mechanical Engineering Publications Limited for
the Institution of Mechanical Engineers, Bury St Edmunds and London, UK.

First Published 1997

ISSN 1356–1448
ISBN 1 86058 098 X

A CIP catalogue record for this book is available from the British Library.

Printed by The Ipswich Book Company, Suffolk, UK.

Contents

Related Titles of Interest

Title	Author	ISBN
An Engineering Archive	D Winterbone	1 86058 053 X
		1 86058 052 1
Train Maintenance Tomorrow ... *and Beyond*	IMechE Conference 1997–1	1 86058 095 5
Automotive Fuels for the 21st Century	IMechE Seminar 1997–1	1 86058 105 6
Energy for the 21st Century	IMechE Seminar 1996–3	1 86058 035 1
Competitive Automation: New Frontiers, *New Opportunities*		1 86058 000 9
Managing Enterprises – *Stakeholders, Engineering,* *Logistics and Achievement*		1 86058 066 1

For the full range of titles published by MEP contact:

Sales Department
Mechanical Engineering Publications Limited
Northgate Avenue
Bury St Edmunds
Suffolk
IP32 6BW
UK

Tel: 01284 724384
Fax: 01284 718692

Preface

In celebrating the 150th Anniversary of the Institution of Mechanical Engineers, we are looking particularly to the future. The symposium *Visions of Tomorrow* brings together key individuals from business and technology to show how engineering will affect our lives in the future.

Increasingly, engineering must be seen in this wider social context, to ensure the right decisions are made, and also to promote engineering as an exciting career with many facets to it. This volume sets out the key issues for the future, as reflected in papers for the symposium held 7–8 July 1997 at the Queen Elizabeth II Conference Centre, London. We are indebted to all those who have contributed to *Visions of Tomorrow*, and those who have joined us at the Symposium. We hope you will find these transactions fascinating reading.

Richard A Pike
Director General of the Institution of Mechanical Engineers

A VISION FOR THE FUTURE

ansport and the community

R ANELL

or Vice President, Head of Corporate Communications and Corporate Affairs, AB Volvo, Göteborg, Sweden

Demand for transport services in OECD countries has for a long time increased at the same pace as GDP. In the emerging industrial economies in Asia and Latin America the growth rate of the transport sector is some 50 percent above that of national income.

Cars, buses and trucks account for a predominant share of transport services in the EU (see pictures 1 and 2). The reason for this can be summed up in one word - flexibility. People are apparently willing to pay for the ability to move in comfort and privacy from one place to another. Companies are forced to reduce capital and shorten lead times. The answer is just-in-time deliveries from door-to-door.

Relative shares differ between countries and regions. Trains are more important in Japan than in Europe. Air travel accounts for 15 percent of person kilometres in the US. But the trend is always the same - towards an increasing share for flexible modes of transport.

The situation in the former Soviet Union differs sharply from the rest of the world. Russia has only one fiftieth as much road mileage per capita as Western Europe. Only 24.000 miles of Russia's roads - less than 10 percent of the total - are highways. As a consequence, in 1990, only 1 percent of freight ton miles went by road as against 33 percent in the US and 70 percent in Western Europe.

Two questions present themselves: will demand for transport services - for goods and people - continue to increase? Will the trend towards more flexible modes of transport continue?

If we consider the global level there is no doubt that demand for transport services will increase at least as rapidly as GDP for the foreseeable future. Some factors will check the expansion of the transport sector, in particular in the mature economies in North America and

Western Europe. The viable size of many production plants can be reduced which may lead to less need for transport. Transmission of digital information will replace transportation of printed material. The general trend towards more "intelligent" production on physical input, less knowledge intensity - reduces the demand for truck transports. A more flexible working life may reduce personal travelling but it is more likely that it will change the way we travel.

However, there are strong forces sustaining the demand for transport services. Whenever the world experiences a period of stability, trade tends to increase faster than production. Travelling is a superior higher good which means that demand increases faster than disposable income.

At present there is no indication that the demand for transport services will be significantly reduced. Our best guess is that it will continue to keep pace with GDP in OECD-countries and expand at a faster pace in the rest of the world.

It also seems likely that the trend towards more flexible modes of transport will continue. One reason is that there is limited genuine competition as far as transportation of goods is concerned. The mode of transport is more or less decided by two factors: distance and value per unit of weight (picture 3). In a society characterized by rapid changes and more flexible life styles, transport services become more heterogenous - it becomes more difficult to predict what or whom shall be transported where and when. Lean production based on reliable just-in-time delivery becomes increasingly important.

Today cars, trucks and buses represent the cost effective solution. As a matter of fact the more than 100 years old inventions of Nicolaus Otto and Rudolf Diesel are still without competition when it comes to efficiency and flexibility.

It is important to add that they represent the solutions to today's demand. It is highly unlikely that tomorrow's transport system can be based on fossil fuel. That is the first and probably most important challenge of the future. A second challenge is related to space and congestion. A third related challenge deals with ways and means to improve the efficiency of the existing infrastructure and traffic system.

Road transports account for some 35 percent of all oil consumption. If one adds shipping and air travel the share will of course be significantly higher. Today we have less than 600 million vehicles - the forecast for the year 2010 is that there will be more than 800 million.

The main emissions from petrol and diesel engines are carbonmonoxide, carbon dioxide, hydrocarbons, particles, nitrous oxides, and sulphur oxides. Some of these emissions have only local effects - others are global and require international cooperation. Some emissions create health problems - others also have an impact on the environment.

There are technical and cost effective solutions to the emissions of carbon monoxides (CO), hydrocarbons (THC), nitrous oxides NO_x) and sulphur oxides (SOx) The modern three way catalytic converter with a lambda sond - introduced by Volvo in 1976 - very effectively takes care of CO, THC and NO_x for petrol engines. A heated converter reduces emissions of CO and THC with some 95 per cent NO_x -emissions from diesel engines are still a significant problem.

However, it is evident, at least in Europe, that the main concern is emissions of CO_2 which are proportional to the burning of fossil fuel.

CO_2 is still very rare in the atmosphere but it has increased since 1957 when reliable measurements started. It is not the only but the most important of the greenhouse gases that may affect the future climate of the globe. There is no full scientific consensus whether the global temperature will increase and, if so, what the consequences would be. However, by way of precaution it is necessary to make every endeavour to reduce the emissions of CO_2.

Naturally the transport sector will have to play its part. It is, however, important to note that emissions of CO_2, wherever they take place, have exactly the same effect on the global climate. This leads to an obvious conclusion. Given the fact that resources are limited we should make investments in CO_2-reduction - which is the same as reduced consumption of fossil fuel - where we get "the biggest bang for the buck". Some studies indicate that the same dollar could by a thousandfold more reduction in India and China than in Western Europe.

Fuel per freight mile has been reduced by some 60 percent between 1970 and 1995. Many factors have contributed - improved engine efficiency, less road friction, less air resistance, increased load capacity, lighter trucks and improved logistics. This development will continue. There is, for instance no doubt that modern technology will improve engines and drive trains. Deregulation in Europe will improve utilization of existing load capacity. Better co-ordination between trucks, ships and train has the potential to yield significant results. Fuel consumption for cars will be reduced because of demand from both consumers and legislators.

But even if the results are impressive we can not avoid the fact that oil is a finite resource. Alternative fuels will remain high on the agenda.

This is the fuel menu studied today by almost all automotive companies:
Petrol
Diesel (Indirect and Direct injection)
Esters and ethers (rape seed methylester, dimethylether)
Gas (liquefied pressured gas, compressed natural gas, biogas, hydrogen)
Alcohol (ethanol, methanol)
Electricity (only batteries or hybrid systems)
Fuel cells

The electric car deserves a chapter of its own. Today this is a car that primarily carries batteries and not much more and not very far. Its future is tied to the technology to store and generate electricity - and a battery is probably not the answer. There is a growing consensus that the most popular electric-drive vehicles will be hybrids - propelled by electric motors but ultimately powered by small internal combustion engines that charge batteries, capacitors or other power sources.

The average power required for highway driving is only about 10 kilowatts for a typical passenger car, so the engine can be quite small: the storage cells charge during periods of minimal output and discharge rapidly for acceleration. Internal-combustion engines can reach efficiencies as high as 40 percent if operated at a constant speed, and so the overall efficiency of a hybrid vehicle can be even better than that of a pure electric drive.

Perhaps the most promising option involves fuel cells. Many researches see them as the most likely successor to the internal-combustion engine, and they are a centrepiece of the ongoing Partnership for a New Generation of Vehicles, a collaboration between the federal government and the Big Three automakers in the US.

However, today the fuel cell is far from being competitive with the cost of internal combustion engines.

Hybrid vehicles will no doubt play an increasing and significant role sometime during the next century. However, if we look at the coming decades we are stuck with the fact that the inventions of Nicolaus Otto and Rudolph Diesel will power some 800 million vehicles.

One thing is clear. There is no perfect, clean and cost effective *substitute* for gasoline and diesel oil when we look at the fuel lifecycle which includes exploitation, refining, distribution and operation (picture 4).

Gasoline, diesel and natural gas score badly in respect of CO_2 and greenhouse gases, but are strong on energy consumption. Biogas (which from a chemical point of view is the same methane as natural gas but is produced from biological material) is rather strong in all areas.

The least energy efficient fuel is ethanol from grain. This underscores a rather general point: the closer to the energy raw material source you can generate your power, the better is the overall efficiency and the lower are the emissions. It would for instance be much more efficient to fuel a power plant directly with the grains than converting them into ethanol.

Looking at the whole spectrum one would probably give three cheers for biogas or natural gas - and put a big question mark behind ethanol.

The second challenge regarded congestion - will there be place for 8 billion people and more than 1 billion vehicles in 2020. How shall we organise transports in megacities with 20-30 million people?

Volvo stated a long time ago that the private car cannot be the primary solutions for flexible mobility in megacities or vast metropolitan areas. If we tried that solution, the car would very soon be neither flexible nor mobile, but simply stuck is a permanent traffic jam. We also realise that the "easy" solution - simply building more roads, highways and parking lots - is not a viable option.

A solution must be based on a smarter infrastructure, price incentives, improved mass transit systems and synchronisation of different modes of transport.

We already have cars, trucks and buses that can receive information about traffic congestion, accidents or available parking space. We need an infrastructure and a logistic system that make use of the smart vehicles. By using devices that sense the distance to other vehicles it is for instance possible to use highways much more effectively.

Modern information technology allows us to charge people who want to use limited space during peak hours. In Oslo an automatic toll generates funding for a dramatic improvement of the traffic system. Price incentives in combination with more flexible life styles should help to distribute the utilisation of space more evenly.

It is evident that a modern city must have an effective mass transit systems. Underground transport is very attractive - but it is very expensive. And even when it is very effective - as in London and Paris - it has to be combined with buses and private cars.

Volvo has developed a mass transit system based on biarticulated buses that carry as many as 220 people. This solution - with a capacity almost as great as an underground system but costing only one fifth - has been implemented in the Brazilian town Curitiba.

Co-ordination of different modes of transport is clearly related to the third challenge - how we can improve the existing traffic system and infrastructure. This will require co-operation among all producers and users of transport equipment as well as the fuel industry.

Consumers can play an important role both as purchasers of vehicles and as voters. There are green consumers willing to pay in order to make their own individual contribution. But I am afraid that we will not see a majority of consumers willing to pay significantly more for the same performance.

The oil industry should be able to provide us with cleaner fuels and cleaner extraction. We are not impressed with the demands on the oil industry in EU's proposed Auto-Oil directive. In Sweden and Finland we already have far cleaner fuels than the directive requires in the future.

Governments have a decisive influence on the price and accessibility of alternative fuels. Already today taxes or subsidies significantly influence the relative price of petrol, diesel and other fuels.

The development of infrastructure is largely in the hands of public authorities. Better logistical support would increase the potential of smart cars. Even more important is to synchronise co-operation between ships, trains and trucks. Such a system would have to be based on standardised lengths and weight for trailers, standardised containers, a logistical system for effective reloading and a supportive infrastructure. Volvo has proposed and developed such a system for Europe but the decision obviously rests with the European authorities in Brussels.

Environmental standards are obviously the prerogative of the legislator. What I would offer here is some modest advice:

- try to establish international or at least regional standards. A proliferation of national standards inevitably leads to less competition and higher costs (not necessarily a problem for the car industry but for consumers).

- green taxes are probably necessary - but avoid de facto discrimination. The eco-tax on virgin fibres proposed in Belgium was discriminatory since the Belgian industry is based on recycled paper. The tax therefore would have entailed a de facto discrimination of imports.

- evaluate before next step. Environmental problems are extremely complex. The models used to analyse global warming are so complex that they are tested against common sense which means that you come very close to adapting the model so that it predicts what you expected in the first place (or even worse - predict what you want).

The best basis for a decision is a thorough empirical evaluation of the effects of earlier measures. In almost all major cities in Western Europe and the US we now see huge improvements of air quality. The likely reason in the extended use of catalytic converters and cleaner diesel engines.

Get your priorities right. The effect of CO_2 emissions is the same wherever it takes place but the cost of dealing with the problem differs radically. The same amount of money would buy a thousandfold reduction in China or India as compared to Sweden according to some estimates. Without vouching for the exactness of these calculations it is fairly clear that there is much to be gained from a cost effective use of limited resources or to put it differently - we must find a way to direct scarce resources for investments where we get the biggest global return.

In the US emission quotas have been made tradable with very encouraging results - maybe this is a possible solution at the international level. The huge differences of costs for the reduction of emissions of CO_2 establish an enormous market with substantial potential gains for both seller and buyer.

CO_2 is a global problem - it demands global action and we should not despair. The Montreal Protocol (1989) for the protection of the ozone layer is an example of effective and prompt international action to deal with a global problem. Lack of international co-operation may actually render national action counterproductive. In Sweden we have proposals to increase CO_2 taxes and shut down "safe" nuclear plants. The result could be to move production to areas where the emissions are far greater. The emissions of sulphur from the oil shale fired power plant in Narva (Estonia) are equal to the total emissions from the Swedish and Finnish industry.

In any case - there is no solution to the CO_2 - problem if there is not an international solution. There is no solution unless it is a common solution.

Fig 1

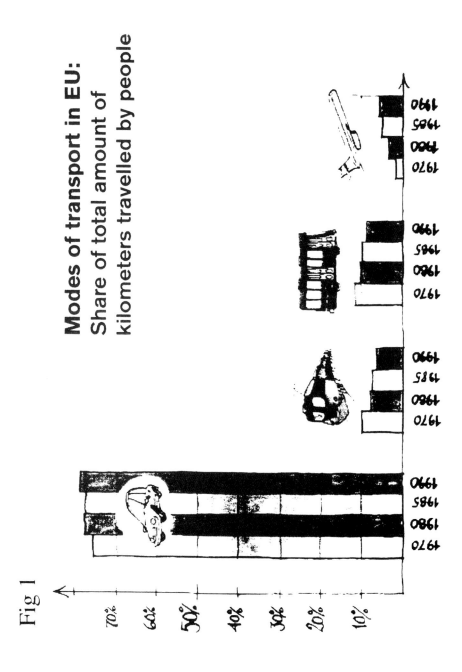

Modes of transport in EU:
Share of total amount of
kilometers travelled by people

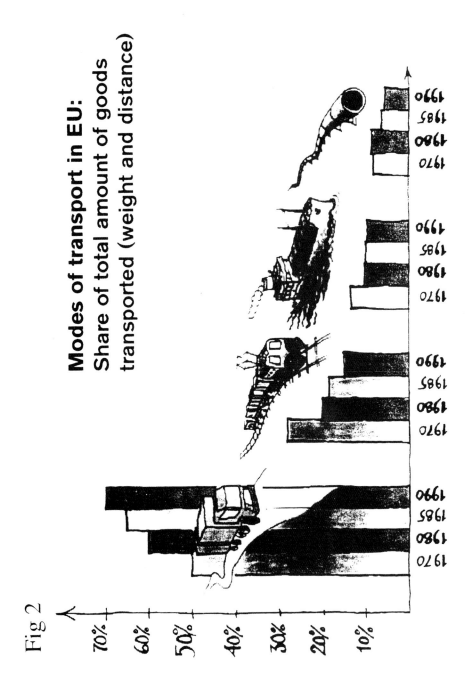

Fig 2

Modes of transport in EU:
Share of total amount of goods transported (weight and distance)

Fig 3
Goods Value and Transport Modes

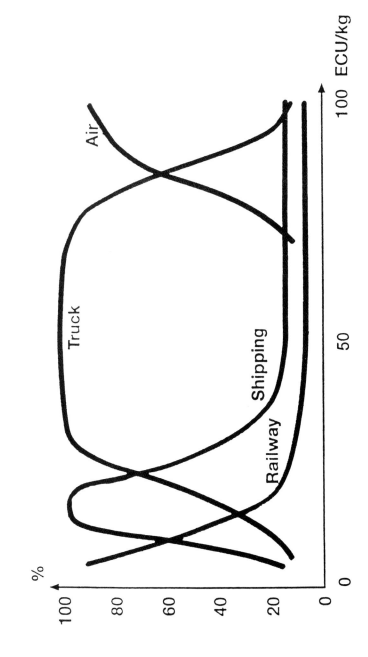

Fig 4

Life-cycle Emission for Different Fuels

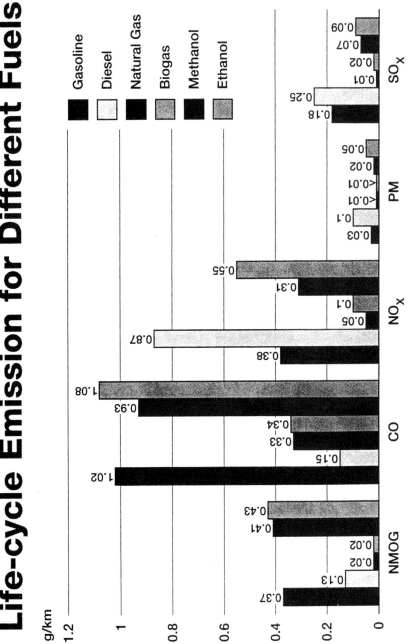

C527/

Manufacturing in the world of tomorrow

D GRANT
Director of Technology, GEC plc, UK

Introduction

In an historical context, manufacturing is a skill that has supported human evolution from its earliest stages. Over hundreds of centuries we have learnt how to transform natural materials into tools, and use these tools and materials to improve the quality of life. The history of manufacturing is well beyond the scope of this paper, however it is useful to briefly note the changing economic and social importance of manufacturing within, say, the lifetime of the Institution of Mechanical Engineers.

In this century many industrialising nations have had huge shifts of people from agriculture into manufacturing, as agriculture became more efficient and as world-wide demand for manufactured goods increased. In recent decades some of these industrial nations have seen further changes to employment patterns as a result, for example, of manufacturers introducing automation or relocating factories to reduce costs and reach new markets. Furthermore, national policies and infrastructures have greatly influenced manufacturing investment and employment. Within the UK, employment in manufacturing has halved in the last 25 years, although output has risen by 10%. In Germany, employment in manufacturing has declined by 5%, and output has increased by 50%. In Japan, employment in manufacturing has grown by 15% and output has increased by 250%.

Manufacturing industry is a vital part of national economic success. Within the UK, manufactured goods account for 64% of exports and manufacturing represents 23% of GDP. Furthermore, by employing 4 million people or 20% of the UK workforce, manufacturing has a strong influence on society as well as the economy. There are 130,000 manufacturing companies in the UK and in addition to their 4 million full-time employees, they support a further 1 to 2 million people in service industries.

International comparisons show considerable variations in the social, economic and political importance of manufacturing industry. A government's influence varies from complete ownership and control of some industries in some countries, through to other countries that influence industry largely through taxation. There have been considerable shifts in these patterns of ownership and influence in recent years; driven to some extent by the liberalisation of markets and internationalisation of companies. These changing and complex interrelationships lead to competition between businesses that involves far more factors than product technology or price. For example, manufacturing companies competing for large international contracts may have very different motives and success criteria. Some may be driven to maintain employment in their home base; some may need to maintain an economic and political alliance; some may wish to develop a technological partnership and some may be driven only to satisfy shareholder financial criteria. Thus, although it is convenient to consider competitiveness between manufacturing enterprises largely through technology, product value and financial measures, the full picture can be much more complex.

The economic, social and political importance of manufacturing industry causes most nations to regularly review the policies and infrastructures that support local and multinational companies. A crucial success factor is often the relationship between national science, technology and education assets and business enterprises. Some nations regularly analyse the potential business and social benefits from technology and market developments. They take long-term *Foresight* visions that can jointly help society, industry, academia and government. This paper is based upon some of the work and findings of the UK Foresight Programme.

The Foresight Programme

The UK Technology Foresight Programme commenced in 1993 as an initiative from the White Paper on Science, Engineering and Technology - "Realising Our Potential". The Programme is managed by the Office of Science and Technology, OST, now within the Department of Trade and Industry, DTI. Sixteen Panels of people largely drawn from business and academe perform studies in sectors addressing groups of markets or technologies. Each of the panels is examining trends and the broad range of forces influencing the long term future of their study areas. These *Foresight* studies are different to *forecast* methods in that they describe a wide range of possible outcomes and scenarios. The objectives are to seek opportunities for improving "Wealth Creation" and "Quality of Life" in the next two decades.

The Manufacturing, Production and Business Processes Panel comprises 25 members drawn from industry and academe. The Panel has a particularly wide study scope and has taken advice from a large number of trade and professional associations, research groups, companies and individuals. These studies of the wide manufacturing base draw upon, and add to, the focused analysis of panels addressing specific sectors such as IT, Chemicals and Materials.

Foresight studies commenced in April 1994 and each panel produced a report in April 1995 that described the issues, the opportunities and recommendations for action. All Panels have continued their work by disseminating the results of their studies and by exploring some topics in further detail. The Manufacturing, Production and Business Processes Panel have "Focus Groups" examining aspects of Business Processes and three manufacturing sectors in greater depth - Machinery Manufacture, Electronics Manufacture and Process Industry.

An important achievement of the national programme has been its broad scope and the direct involvement of over 10,000 people from a wide range of disciplines. The programme is delivering three types of benefit. Firstly a set of visions, forecasts and scenarios covering social, economic, and technological possibilities in many subject areas over the next two decades. Secondly, the programme is generating recommendations for changes that will improve UK social and economic conditions. Finally, the *Foresight Process* itself is demonstrating the added value of exploring future issues through discussions, studies and networks involving people with differing knowledge, skills and perspectives.

The future success of UK-based manufacturing industry will be influenced by the Foresight Programme. A large number of recommendations for change have been made, and some projects are now under way to strengthen the technology base in subjects that will enhance industrial competitiveness. Examples include new materials, new processes and innovative Information

Technology applications. Foresight Programme recommendations also propose changes to education and training; covering both educational content and delivery. Another potentially important influence on manufacturers is the encouragement and guidance to perform Foresight studies of their own. This should improve the quality of their forecasting and strategic planning exercises, and should also reveal competitive business processes and innovative product and service opportunities for emerging markets.

The objective of this paper is to explore some of the many factors that will influence the future success of manufacturing industry; to present some scenarios set in 2015, and to outline some of the influences on - and opportunities for - UK based manufacturers. The paper discusses studies performed in the UK Foresight Programme and the author gratefully acknowledges the excellent work of all Panels, and in particular the innovative views from colleagues on the Manufacturing, Production and Business Processes Panel.

History and Hindsight
To gain an initial impression of the potential magnitude or impact of changes in the next two decades it is helpful to briefly consider recent decades during which UK-based manufacturing has undergone considerable change. This change has left a smaller manufacturing base, but part of that base is leaner, more efficient, and potentially capable of seizing future opportunities. It is a matter of continuing analysis to determine just how much of the manufacturing base has improved and some studies describe a long "tail" of less efficient businesses whose poor performance statistics mask improvements to the main "body" of manufacturing industry.

Today there are approximately 130,000 manufacturing companies in the UK with a total of four million full-time employees. Almost 30% of UK manufactured output is exported and this accounts for 64% of all UK exports. Manufacturing represents 23% of GDP and directly employs almost 20% of the workforce. It is reported that more than one million jobs in the service sector are directly dependent upon the manufacturing base, and a further three million are indirectly dependent. The largest manufacturing sectors are: Food & Drink; Electronics & Electrical Engineering; Paper, Printing and Publishing; Chemicals and Pharmaceuticals; and Mechanical Engineering. One quarter of UK-based manufacturing industry is foreign owned.

Between 1960 and the 1990s, UK Manufacturing output has marginally increased, whereas that in Germany has doubled, in France has tripled and in Italy has quadrupled. Japan's output has increased tenfold. The UK share of world exports at around 6% has reduced to one third of its former strength, whilst Japan has seen a threefold increase. The number of full-time employees in UK manufacturing has halved since 1970. The reduction of 30% in the 1980s compares with reductions of 17% in France, 11% in Italy, 5% in the USA, no change in Germany and an increase of 13% in Japan. Analysis of labour productivity shows that although the UK made considerable improvements in the 1980s, it still lags well below France, Japan and the USA with a gap of 25% to 40%. Investment inside factories and in Research & Development, despite many strengths in the national science and technology base, are also reported to be below that in many other nations.

There is, of course, considerable discussion and analysis of these aggregate statistics. Some UK industry sectors and some companies perform well above the average level. Unfortunately many other companies still perform well below levels necessary to sustain competitiveness and growth.

In summary, the trends and statistics reveal considerable changes over recent decades. Declining numbers of people are directly employed in manufacturing; there is increased manufacturing productivity but in aggregate it is still below that in other countries; production for domestic consumption has been rising more slowly than domestic expenditure on manufactures; in aggregate, the UK invests less in manufacturing than comparable nations; the UK hosts a growing proportion of foreign owned businesses (and has increased overseas investment by UK enterprises); and much of UK-based industry faces strengthening competition that is influenced more by international markets and companies than by local players.

Yet manufacturing strength remains vital to the future social and economic needs of the UK, so how can we describe a potentially successful future and then influence the future to our benefit?

The UK Foresight Process
The programme and methodology used by all Foresight Panels is broadly similar. The Manufacturing, Production and Business Processes Panel commenced in April 1994 by collecting and analysing advice on the *Issues*, *Trends* and *Drivers of Change* across a wide spectrum of industries. Further consultation, regional workshops, expert sub-groups and detailed analysis enabled the Panel to identify potential scenarios for success in 2015; and priorities for near-term actions to improve the probability of long-term success.

The Manufacturing Panel identified a very large number of forces or *"drivers of change"* that influence the future success of UK-based manufacturers. The forces were categorised under five headings or "STEEP" factors - comprising Social, Technological, Economic, Environmental and Political change. The more important forces from these lists, some independent and some acting together, were selected for further analysis. These factors helped the Panel develop a range of scenarios for UK-based manufacturing in 2015.

Foresight in Manufacturing
Six of the "drivers of change" that will influence the future success of many manufacturing businesses are described below. The six generic topics will in practice be associated with specific market and technology developments that each manufacturer employs to gain competitive advantage.

1. Internationalisation - the influence of overseas markets and enterprises.
2. Changing customer requirements and sales & service networks.
3. Competitiveness through agility, focus and partnerships.
4. People skills, motivation, leadership, teamwork, learning, knowledge and innovation.
5. Responses to environmental concerns and regulation.
6. Influence of Information Technology and communications.

1. Internationalisation
Throughout the world, the basic resources of manufacturing are becoming more widespread and more mobile. Factories, equipment, supplies, people, technology and finance regularly cross borders that once protected local trade or appeared to offer few market opportunities. Furthermore, the growing international telecommunications and transportation networks support and encourage the wider distribution of trade and manufacturing.

Emerging economies, many with highly skilled people, are rapidly developing low cost manufacturing operations to satisfy local needs and to capture both local and export business from more costly suppliers. The world-wide availability and diffusion of technology is enabling countries with good educational standards, skills and strategic partnerships to rapidly develop high technology businesses. The newly developing and industrialising economies represent a combination of "opportunity" and "threat" to established exporters from the UK. As a market, they require increasingly advanced goods. As a supplier, they can swiftly compete with advancing products at low cost. Clearly, one of the more important requirements for large and small UK enterprises is to develop strategic plans for relationships within developing countries.

Multi-national enterprises, MNEs, will continue to be in a strong position through their ability to rapidly locate selected parts of an enterprise into a new area, often creating local partnerships and supply networks. Success requires excellent local knowledge and links. Some entrepreneurial UK small and medium enterprises, SMEs, globally partner the larger MNEs so that they benefit from the MNE expansion.

One consequence of the strength and global outlook of MNEs will be the reducing loyalty to, and dependence upon, a single or base nation. The giant MNEs of the future will adapt to local conditions, and be linked to their founding organisation through the strength of internationally recognised brand names and protected technology sources. Technology development will be performed internationally to achieve best value for money as this will often be a substantial business cost. Key competitive strengths will involve process skills capable of rapidly adapting and implanting an entire business into a new country. Technology acquisition, management and transfer will be a related key skill. It is probable that the MNE will focus internally on a manageable set of competencies and supplement this set with a range of international partnerships for "non-core" content.

It is interesting to consider the relative strengths, roles and interactions of the large MNEs and the government of host nations. In some countries an MNE will make little long-term commitment, whereas in other countries they will develop stronger bonds through, for example, local or regional R&D activities. Although the strategies likely to be adopted by MNEs and governments are difficult to predict, it is likely that some MNEs will yield considerable international power.

A form of business alliance, the "virtual enterprise" business model, will play an increasingly important international role. In essence, the model comprises many small and specialist enterprises that connect in a network to satisfy a particular need, and upon completion they may disperse to form other networks. It is possible that at any point in time an enterprise will be associated with a number of networks. The model is particularly suitable for international trade, and will be facilitated by the growing power of telecommunications networks. The Foresight studies highlight the current UK competitive advantage of English Language and a good telecommunications infrastructure. Continued infrastructure development, together with an "IT literate" population with wide international knowledge and contacts, could generate huge business opportunities by the 2010s. Specialist and high technology skills will be in great international demand, and UK-based networks could achieve a leading role.

2. Changing Customer Requirements

Demographic and social changes are rapidly changing many markets. The growing levels of international consumerism are creating demand for previously unavailable, unaffordable or non-essential goods. For example, ageing populations in many countries focus demand on healthcare products, whereas the affluent young in other nations focus demand towards personalised and leisure products.

World population growth of 80 million per annum stretches existing resources at a time when global sustainability is a widespread concern. Food, water and electrical power are basic needs in developing countries but are also influencing markets in developed economies. Domestic products, for example, are increasingly required to perform their function with considerably less energy and water. Furthermore the environmental pressures discussed later greatly influence consumer demands and legislative controls.

Foresight studies of many markets suggest the possibility of huge changes in market opportunities over the next two decades. Customers will increasingly seek products that form part of a total service offering. Value and quality will be judged by the supplier's ability to satisfy the total requirement. Most manufacturing businesses will choose either to respond to these changes or, perhaps wisely, will anticipate future needs and opportunities, and develop new products and services before their less visionary competitors.

Sales and customer support networks will change to incorporate more Information Technology and Communications products and services. In some cases this will put the customer directly in contact with the manufacturer. It will potentially give far greater power to manufacturers, and help extend that power internationally. Some of today's retail sales and service networks will either decline or change. Some manufacturers may choose to move higher up the value chain into service and support activities, and rely upon specialists and low cost factories to perform more of the basic manufacturing operations.

3. Competitiveness Through Agility, Focus and Partnerships

In the next two decades most manufacturing sectors will need to respond to rapid market changes and increasingly rapid competitors. This will require rapid process and supply chain response times and also require flexibility to meet changing product volumes and mix. These requirements, often accompanied by increasing product and process complexity, will cause some manufacturers to focus on a more limited set of tasks than today and to develop partnerships that satisfy their "non-core" needs.

Observations from the huge international effort currently devoted to business process analysis suggest that process agility is a particularly difficult objective. This is especially so when "reengineering" businesses built up over a large number of years. One challenge, for example, is that many current businesses rely on IT systems and networks with limited flexibility. These systems have often automated a complex set of interdependent processes, but are intolerant to substantial process change. This constraint requires a new approach to IT if business "agility" is to be achieved in future years. Clearly, the requirement (and opportunity) is to deploy *agile business systems*.

The "virtual enterprise" model is an effective way of achieving agility through the creation of networks of units, each with a specialist focus. A vision of 2015 suggests that virtual enterprises will capture a considerable proportion of many markets and that IT will by that time operate seamlessly across many enterprises and processes throughout the product life cycle.

Successful managers of business and manufacturing processes who are improving "quality" have already extended their influence from the narrow focus on products to much broader management through TQM, Total Quality Management, of quality throughout the enterprise. It is expected that this will expand further to encompass more of the interrelationships with society and give a wider consideration of the value supplied to customers. The most successful businesses will continue to be those with total control over an integrated set of effective processes that fully satisfies their customers and stakeholders.

In many industries it is likely that the current concept of a "supply chain" will evolve into integrated partnerships that optimise processes across many enterprises. Teamwork and other principles that prove successful in one enterprise will be extended to all enterprises involved in the partnership. Project leadership will be given to those who have greatest ability to manage a changing network of enterprises and skills throughout the project or product life-cycle.

4. People skills

The competitiveness of manufacturers is largely dependent upon the skills employed throughout an enterprise and its associated partners. These skills are created and influenced by the national educational system and by subsequent training. Motivation and culture form an important part of this complex picture. Looking ahead at the increasing pace of international development it is clear that education and training must become a continuous process setting ever higher standards for all to achieve.

Emerging economies have placed great importance on, and made impressive progress towards, the creation of well educated and trained people at every level of their new industries. The longer established economies can compete only through renewed attention to skill development. This must be accompanied by national infrastructures that encourage scientific and technological progress thus equipping companies with staff educated with internationally competitive skills and knowledge.

Foresight studies and many other national and international projects are examining the future requirements for education and training. Perhaps it would be appropriate to comment on just a few of the many related issues addressed by the Manufacturing Panel. The Panel highlight the need for the development of teamwork, communication, interpersonal skills, problem solving, international business awareness and business *process* awareness. Particular emphasis is placed on the need to relate many industrially relevant specialised skills, such as mechanical engineering, to a wider knowledge of business processes in an enterprise. Specialist technical skills will continue to be very important, but it will be beneficial for the specialist to appreciate customer needs and the enterprise's processes for developing markets or responding to opportunities. This broad business process knowledge will enable the specialist to select the most effective or innovative method of applying the skill.

One of the specific opportunities addressed by the Panel was the possibility of taking more training directly to the employee or manager at their place of work. The growing power of IT and communications networks will enable this service provision.

Leadership skills require increased emphasis on the management of change and the development of responsive, flexible teams and organisations. Virtual enterprises and strategic partnerships will require considerable reliance on trust, teamwork, co-ordination and communication - often across great distances and cultures.

It is important to note that for a variety of reasons employees move from company to company. In 1994, it was reported that nearly 20% of UK employees had been with their employer for less than a year, and 50% had been with their employer for less than five years. For some people, job-hopping or even career-hopping will become part of their life. Consequently, employers will take actions that encourage loyalty where necessary, but will also have to ensure that their processes are tolerant to employee moves and ideally will benefit from them.

5. Responses to environmental concerns and regulation

Concern for the environment has become a major public issue in industrialised countries in recent years. Public pressures are persuading governments to introduce legislation. Its influence extends to most aspects of manufacturing. Manufacturers must take responsibility for the life-cycle of their products; and for materials, waste, energy and people directly or indirectly involved with the production process.

Products and the production process will increasingly be influenced by recycling, reuse and disposal issues. Although there appears to be little pressure at present to design and manufacture products for a substantially longer life, it may well become a requirement in some sectors over the next two decades.

Environmental issues will influence most production processes in current use. Foresight studies of the electronics industry, for example, envisage considerable changes as new materials and processes are developed to replace lead-based solders. Similar massive scale changes are already being introduced as CFCs are eliminated. Whilst changes represent a threat to existing processes, they also represent a huge opportunity to those who develop or implement clean technologies. There will be a large market for many products that improve energy efficiency, reduce pollution and waste and allow the reuse or safe disposal of materials.

A wide range of design and manufacturing processes will change to minimise energy and waste. Current trends towards "design for assembly" will encompass longer term "design for disassembly" requirements. New manufacturing processes are envisaged for producing products without wasteful and expensive machining operations. Some of these processes may, for example, develop from currently emerging "rapid prototyping" technologies.

6. Influence of Information Technology and Communications

IT and communications technology developments will continue to be one of the most influential drivers of change in the manufacturing industry. Manufacturers already benefit from IT applications in design, production and many business process. Former "islands" or "threads" of IT are increasingly being integrated within an enterprise. Systems are extending into the supply chain and are directly linking to customers.

The full power of IT and communications requires Open Systems with compatibility, connectivity, upgrade paths and international standards. These are very demanding requirements because of the rapid advance of underlying electronics technology. Computing power is likely to continue reducing in cost by 30% every year, and performance power will double every 18 months, for at least the next decade. It is therefore envisaged that the power of this key enabling technology will make huge advances, limited only by a lack of imagination or a lack of development resources.

The increasingly important Virtual Enterprise business model requires an underlying IT and communications network. The attributes of speed, spread, flexibility and service content must be supported by high levels of security and privacy. Many enterprises, will employ "Virtual Manufacturing" technologies to simulate and model each product and the related production and business processes. These models, and related "Synthetic Environments" will become key asset:

for any enterprise. It will become possible to model with increasing accuracy the product and its life-cycle processes, from raw materials through production and use to its eventual disposal.

A key trend in all manufacturing businesses, and especially in electronics, is the increasing complexity of products and associated processes. "Product data management" systems are being developed to assist with this need, however these systems must evolve to support an entire enterprise. The management of **complexity** will, in the author's opinion, become a key determinant of future business success.

2015 SCENARIOS

The "drivers of change" described above, together with other "STEEP" influences, were used by the Manufacturing Panel to envisage national or business enterprise "scenarios" set at some future time. This was an important output of the foresight process because it brought together a range of visions, some of which appeared incompatible or carried divergent opinions. The exercise aimed to provide a diverse set of scenarios where the feasibility and desirability of each could be explored. The particular scenarios outlined by the Manufacturing Panel of the Foresight Programme represent possible, but extreme, modes of success for UK manufacturing. Two of the scenarios are described below:

The Panel looked at the successes possible from a UK base dominated by Design/Marketing/Business Management ("Design-Shop") activities, and contrasted them with a UK dominated by Production/Manufacturing/Assembly ("Manufacturing-Shop") activities.

UK "Design-Shop"

The Design Shop scenario describes the UK in 2015 as a country with excellent knowledge of global markets, their specific trends and future requirements. A highly educated and extremely innovative workforce would identify product opportunities throughout the world, specify and design the products for low cost manufacturing elsewhere, and work with the selling, service, financial and other support groups necessary to make the product a success in the target markets. These leading-edge "design shop" companies would be supported by a world-class science and technology base; and related skills and activities such as rapid prototype developers, IT system developers, simulation and design tool developers. Communication networks would span the globe, and working patterns would fall into line with the requirements of each opportunity as it arises. A strong financial services sector would support this global network. Entrepreneurs would flourish throughout this rapidly growing economy.

UK "Manufacturing-Shop"

The Manufacturing Shop scenario describes the UK in 2015 as a country with many high technology, high skill, high quality manufacturers with their supporting infrastructure of machine tool and automation systems companies. The scenario considered that the country could not compete with low wage economies for the low technology end of the product spectrum, but had chosen instead to become Europe's leader in high technology manufacturing - giving high added value. A responsive, flexible, motivated and well-trained workforce operate, and automate where necessary, the most efficient and environmentally clean manufacturing operations in their chosen sectors. A strong supporting national infrastructure would include excellent transportation systems and fast routes to the rest of Europe. A thriving services sector would support the manufacturing companies. Closely coupled local supplier networks provide specialist materials and equipment. Government planning for education, taxation and infrastructure was a key to success in this scenario.

Recommendations for Change
The trends, visions and scenarios outlined above identify some of the challenges and opportunities ahead. From this study and much additional analysis, the Foresight Panels are recommending many changes considered necessary to enhance the future success of UK-based manufacturers. Specific recommendations are directed towards government departments with responsibility for industry, for education and for the national science and technology base. Further recommendations directly address industry.

Recommendations from the Manufacturing Panel covered six topic areas:

1. Increased emphasis on business processes (by industry, academe and education)
2. Technology development priorities (in the national science and technology base)
3. Improvements through education and training
4. Extended vision for manufacturing and production businesses (their own *Foresight*)
5. Improved communications and support networks (including local business networks)
6. National infrastructure for improved competitiveness (for Government action)

Government departments and many manufacturing companies are already implementing changes, and manufacturers are continually increasing their efforts to improve their measures of competitiveness. It is therefore necessary to question whether these activities are sufficient to address the future issues and opportunities envisaged in *Foresight* studies, and whether enough UK-based companies are taking necessary steps.

The answer to these difficult questions requires careful comparative analysis and "benchmarking". These comparisons must examine the current situation, the rate of change and future projections for competitors. Benchmarking at a national level provides broad economic indicators and some relevant measurements such as education and innovation in science and technology - although it is difficult to provide exact international comparisons. Benchmarking at an industry or company level provides more specific guidance on the measures necessary to improve competitiveness. These improvements require some actions within the company or industry and some actions at a national level.

A number of recent international benchmarking studies made available to Foresight Panels give cause for concern. These studies indicate that performance gaps between leading companies and their competitors exceed 30% in many measures. These gaps suggest a need for large and possibly radical actions rather than small annual incremental improvements.

Step changes are possible, for example, by extending the role of the manufacturer - as envisaged in some Foresight scenarios - to provide a range of services to a customer that adds additional value to the provision of the manufactured product. However this approach is not always possible, and in most cases it will be necessary to radically change existing processes. These changes will involve many suppliers and partners; and for the longer term, must involve national infrastructures.

Important sources of assistance to manufacturers are the networks, often at a local or regional level, that enable discussions on subjects such as best practice and benchmarking. These networks are clustered around trade and professional associations, universities and regional government departments. These networks have been helpful to the Foresight Programme, and should be encouraged to engage in their own Foresight work.

Conclusions

Success in the manufacturing sector of the economy is vital to future national prosperity. Analysis of a complex international picture suggests that there will be a growing world-wide need for manufactured goods and related services, many of which can be based upon current UK strengths in, for example, the science and technology base. Consequently, the prospects for UK-based manufacturing industry over the next two decades and beyond could be excellent, provided that all stakeholders recognise and plan for the business opportunities but also overcome potential threats or weaknesses. Successful manufacturers will need people with excellent skills, innovative technologies and effective processes and partnerships to create products and services that are valued by their customers.

The national Foresight Programme can not define every action a manufacturer might take to capture future opportunities. What it has done, though, is to express the possible - or probable - scenarios in which successful manufacturers will operate. It has identified many potentially important technologies, and has demonstrated opportunities for beneficial collaboration between industry and those in education, science and technology. One of its greatest achievements has been the development of networks of people with a broad range of skills and perspectives who are able to study and then influence the future.

References

Technology Foresight - Progress Through Partnership
Volume 9 - Manufacturing, Production and Businesses Processes.
April 1995
HMSO
ISBN 0-11-430123-9

ngineering in health and medicine: past contributions nd future opportunities

fessor R M NEREM PhD, MNAE, MIOM, MAPS, MBMES, MIFMBE, MIUPESM, FAAAS, FAIMBE, FASME
ctor, Institute for Bioengineering and Bioscience and Parker H Petit and Chair for Engineering Medicine,
rgia Institute of Technology, USA

1.   INTRODUCTION

On the occasion of the 150th anniversary of the Institution of Mechanical Engineers, it is appropriate to recognize the contributions of engineering to the advancements in health care and medicine and the growing area of medical and biological engineering, or what some call bioengineering. In the last 50 years this field has moved from one in which the more traditional areas of engineering have applied their knowledge to solve problems in medicine and biology to the point where a true biology-based discipline is beginning to emerge.

From a mechanical engineering viewpoint, it is the field of applied mechanics which has led the way. The contributions of biomechanicians, many of whom were or are mechanical engineers, have been wide ranging, and these will be reviewed in the next section. However, as we will see, the impact on the field of mechanical engineering has been much broader than just biomechanics, and there is a medical product industry to which mechanical engineers have made an enormous contribution. Furthermore, with the advent of the biological revolution, bioengineering is changing as will be discussed later.

2.   THE APPLICATION OF THE PRINCIPLES OF ENGINEERING

The contributions of engineers have really been of two major types. First, there are the variety of contributions that have been made which have improved our understanding of how the human body functions. The human body is a fascinating machine, one which integrates virtually every field of engineering endeavor into a single functioning system. Here we will focus on the mechanical aspects of this system, and because my own background is applied mechanics, we will focus on examples from biomechanics (1).

If we take the circulation system as one example, its function is to deliver nutrients and remove waste products. It is a flow system, one with a mechanical pump driven by chemical energetics, but also one where the biology is intrinsically linked to the mechanics. An area of interest which has motivated much of the research on blood flow dynamics is the investigation of the role of hemodynamics in the genesis of atherosclerosis, a major contributor to heart disease (2). In the 1960s there was the debate between the high shear stress theory and the low shear stress theory. Thirty years later there seems to be agreement that in fact early lesions are localized in low shear regions. The question is "what are the fluid dynamic features of a low shear region which are mechanistically involved in the genesis of these lesions, i.e. atherogenesis?"

Another example is the musculo-skeletal system. This system of muscles and bones is designed to carry mechanical loads, while at the same time making human locomotion possible. The area of orthopaedics is one where biomechanics has become totally integrated in clinical medicine. No major orthopaedic department is devoid of the teaching of biomechanics and many have biomechanicians as part of their faculty or staff. Research has focused on both hard tissue, e.g. bone, and soft tissue such as cartilage, ligaments, and tendons. Also included is sports biomechanics, i.e. the application of the principles of mechanics to injuries in sports.

Moving away from applied mechanics as a discipline, another area in which engineers have made a contribution by applying what might be called first principles is bioheat transfer and the application of its principles to cryobiology. Over the past three decades there has been an increasing number of engineering-trained researchers who have made the field of cryobiology a primary focus of their work. One area of focus in cryobiology that engineers have emphasized is the design of new apparatus, including both experimental instrumentation and clinical diagnostic and therapeutic devices. Examples include instruments invented to enable the quantitative control and measurement of the fundamental phenomena that govern processes in cryobiology. Other applications include the development of novel measurement techniques for assessing system properties and states during freezing and thawing. In cryosurgery and in cryopreservation new probes and apparatus have been designed to provide more accurate and effective processes to achieve clinical objectives. Equally important and complementary to the design of hardware is the development of analytical models. Application of these experimental and analytical tools has enabled the optimal design of cryosurgical and cryopreservation protocols.

3. MEDICAL DEVICES AND IMPLANTS

Engineers also have contributed to health care, this through the development of devices, instrumentation, and medical implants. It is the last of these which will be focused on here for the medical implant industry has become a global, very competitive industry, one with annual sales in excess of $50 billion.

One example is the field of rehabilitation technology which had its origins, for the most part, in the field of limb prosthetics. The important clinical successes achieved by mechanical engineers working in the prosthetics field, particularly after the World Wars, demonstrated the benefits that engineering can have on the rehabilitation of persons with impairments and disabilities. The successes associated with provision of mobility and manipulation for persons with amputation are, to great extent, what propelled the field of rehabilitation engineering into existence worldwide. Thus, the current activities on communication aids, computer access, wheelchair design, seating and positioning, functional electrical stimulation, adaptation of the environmental, rehabilitation robots, therapeutic equipment, etc. are all indebted to mechanical engineers who "early on" developed limb prostheses.

One of the best known prostheses in history was a prosthesis made famous by the person who became the Marquis of Anglesey. He was in charge of the cavalry for Wellington at Waterloo. His leg was amputated by a cannonball, one of the last cannon shots of the battle. A beautiful wooden leg was fashioned for him, which contained an novel mechanical principle still in use today in one leg prosthesis, which happens to be sold by a French company no less! The leg, patented by James Potts of London in 1800, became known as the Anglesea Leg. The leg was later brought to America where it was modified. It became one of the prostheses widely used by amputees from the Civil War, and there were thousands. Thus you can be sure that English mechanical engineers contributed importantly to the walking ability of many amputees in my state of Georgia after the war.

It should be noted that it was after World War II, i.e. in the 1950's, that engineering's contribution to medicine took on a markedly different role. Prior to this time, the focus was largely on diagnostic equipment, orthopaedics and rehabilitation (braces and wheelchairs) and on surgical instruments. Development in the '50s of the first large diameter vascular grafts, prosthetic heart valves and the cardiac pacemaker launched the era of medical device-based therapeutics. For the first time technology was able to restore lost or diminished function inside the body. It was able to do this in a way that raised significantly the patients' quality of life. Furthermore, medical device technologies often do this in a more cost efficient manner than prior methods.

4.   THE BIOLOGICAL REVOLUTION: NEW AREAS FOR ENGINEERING RESEARCH

With the advent of the biological revolution, medicine has moved to the cellular and molecular levels. It was simply a matter of time before medical and biological engineers would also move to this microscopic level. Today we have people doing cellular biomechanics, even molecular mechanics, and also cellular engineering (3). There is the International Conference on Cellular Engineering, the 3rd of which will be held in San Remo, Italy in September of this year, and the International Federation of Medical and Biological Engineering has launched a new journal entitled *Cellular Engineering*.

On the basic research side of the field, one example of cellular biomechanics or engineering is the study of the influence of the physical forces imposed by the flowing blood on the structure and function of the vascular endothelium, that monolayer of cells which makes up the inner lining of a blood vessel and thus is in direct contact with the blood flow. It was really engineers who first recognized the importance of the mechanical environment in which cells resided, and this has become an expanding area of activity. Not only is the structure of cells modified by the external, physical force field, but such basic functions as cell proliferation and the synthesis and secretion of biologically active molecules also are influenced. A fascinating question is how does a cell recognize the mechanical environment in which it resides? Also, once this recognition has taken place, how is this signal transduced into the kind of changes which have been observed to take place? These are fascinating questions which now are being actively addressed.

5.   TISSUE ENGINEERING: AN EMERGING TECHNOLOGY

The focus at the cellular and even subcellular, molecular level is also having an influence on industrial applications. An example of this is the emerging technology of tissue engineering. The definition of tissue engineering is the development of biological substitutes, ones using living cells and natural biological materials, at least in part, to implant so as to replace, repair, maintain, or enhance function (4). Tissue engineering is one of the new industries which is based on the science of biology. It is at the interface of the traditional medical implant industry and biotechnology. It represents the next generation of medical implants, and as an industry tissue engineering is just beginning to emerge. Just within the U.S. there are close to 40 companies with activities in tissue engineering, ones which involve several thousands of employees, and industrial research and development expenditures estimated to be in excess of $300 million annually.

There are many opportunities for tissue engineering, with the first products being in the area of skin substitutes. Both Organogenesis (Cambridge, Massachusetts) and Advanced Tissue Sciences (La Jolla, California) expect to have a skin equivalent, one which incorporates living cells, approved by the U.S. Food and Drug Administration during 1997. The development of skin substitutes, however, literally represents just "scratching the surface." At Georgia Tech in our Institute for Bioengineering and Bioscience we have a number of new thrusts as part of our Tissue Engineering Initiative. Much of our focus is on tissue engineered, cell-seeded constructs where the primary function is a biomechanical one. Applications here include cardiovascular substitutes and orthopaedic implants. However, we also are interested in other types of constructs, particularly ones where physical factors may be important. An example is the area of encapsulated cell therapeutic approaches, where the cells may be physically encapsulated in order to achieve immunoisolation. One application of this technology is where cells which are insulin secreting and glucose responsive are encapsulated to create a bioartificial pancreas.

6.   GENETIC ENGINEERING: IS IT REALLY ENGINEERING?

With the advent of recombinant DNA technology, a whole new industry was started, what is sometimes called the biotechnology industry, other times the genetic engineering industry. In general this involves the genetic manipulation of cells using recombinant DNA technology to enhance production, e.g. of a certain protein, so as to manufacture it cost effectively and thus to make it available for wide-spread clinical use. In the development of bioprocessing systems for this, there are many important mechanical engineering problems. There are other applications for the use of the genetic manipulation of cells. This includes gene therapy and also tissue engineering. In regard to the latter, here the motivation is the enhancement of a certain type of cell behavior, e.g. cell adhesion, cell locomotion, or even immuno-suppression.

Now there are some who say that genetic engineering is not engineering! However, if an individual by design genetically manipulates the genes of a cell so as to produce a certain specific type of behavior, why is this not as much engineering as other, more conventional engineering activities? I believe that there will be an increasing group of engineers who will want to be involved in genetic engineering. The issue is no longer simply the genetic manipulation of a cell. Rather, it will be a design goal, e.g. the manipulation of the genes of a cell in such a way so as to design in a certain quantitative relationship between the behavior of a cell and its external environment. If the design goal is one where it is the mechanical function of a cell is being attacked, then perhaps the genetic engineering will be done by a mechanical engineer.

Either individuals trained as engineers take up the challenge, or else this type of engineering will be done by people trained in schools of biology. After all, the reason genetic engineering is viewed by engineers as not being engineering may have nothing to do with whether or not it really is engineering. Our bias may be predicated on the notion that it is only engineering if it is done by people trained as engineers. However, is it possible that this modern day engineering will not be done by engineers because of our woeful lack of knowledge about biology?

7. MOVING INTO THE 21ST CENTURY: IS THERE A ROLE FOR THE MECHANICAL ENGINEER?

As we end this century and enter into the next one, we are moving into a new era and it may be logical to ask the question, "Is there a role for the mechanical engineer in the field of medical and biological engineering?" The simple answer is "yes" for there will continue to be problems where an individual who is well trained in the field of mechanical engineering will be able to make a contribution. For example, the use of rapid prototyping will impact the medical implant industry, not only the more traditional types of products, but even the newly tissue engineered products.

However, the challenges of the 21st century will spawn a new type of medical and biological engineer, a true bioengineer, one who practices a biology-based engineering, an individual who will be able to integrate the biological revolution into engineering. For this new bioengineer, the three letters "bio" will no longer be a prefix for it will be necessary to put equal emphasis on the bio part as on the engineering part of the word. Bioengineering, i.e. biology-based engineering, is emerging as a new discipline. It will become an equal partner in the world of professional engineering with those engineering specialities based on physics and chemistry.

Should the mechanical engineering profession be threatened by this? I think not. There will still be plenty of opportunities for mechanical engineers, and furthermore, many of this new breed of a biology-based engineer will come out of mechanical engineering. In many ways perhaps the closest model is that of materials science and engineering where, in spite of the fact that there are academic departments of materials science and engineering and individuals who wear this "hat," the practice of materials science and engineering is done by a wide variety of people, not the least of which are those from mechanical engineering.

Rather than be threatened, we should enter into what I believe can be a celebration of a wider array of opportunities, new challenges, and perhaps the greatest intellectual diversity ever confronted by the engineering profession. Biology has become too important to be left to biologists. In fact, not only is bioengineering emerging as a separate discipline, but what will come out of this could revolutionize the field of engineering itself.

ACKNOWLEDGMENT

The author appreciates the assistance of the following individuals who are pioneers in their own right in this emerging field of bioengineering: D. Childress, P. Citron, K.R. Diller, R.D. Kamm, W.M. Phillips, A.B. Schultz, and R. Skalak.

REFERENCES

1. Mow, V.C., Nerem, R.M., and Woo, S.L-Y., Special Issue: 20th Anniversary Biomechanics Symposium, ASME J. Biomechanical Engineering, Vol. 115, No.4(b), 1993.

2. Nerem, R.M., "Vascular Fluid Mechanics, the Arterial Wall, and Atherosclerosis," ASME J. Biomechanical Engineering, Vol. 114, No.3, pp.274-282, 1992.

3. Nerem, R.M., "Cellular Engineering," Annals of Biomedical Engineering, Vol.19, pp.529-545, 1991.

4. Nerem, R.M. and Sambanis, A., "Tissue Engineering: From Biology to Biological Substitutes," Tissue Engineering, Vol. 1, No. 1, pp. 3-13, 1995.

formation technology – making the future happen

ARNSHAW BEng, FIEE,Director IEC, President IBTE
naging Director, BT Network and Systems, BT Centre, London, UK

INTRODUCTION

For some "the future is now", a phrase which in a way is a compliment to the engineering profession and which in itself reflects the pace of change. However it also signals complacency, for continuous improvement and innovation, making the future happen is what we, collectively, are all about. The fact is, technology is propelling an unprecedented change in society. More than ever before the engineer needs to understand the social and organisational implications of the professions contribution to society.

The business of the engineer is indeed the future. I would like to review how changes in society, many of them triggered by technology, point us towards redefining the role of the engineer.

UNDERSTANDING THE PAST

For telecommunications businesses, the past was very different.

Above all, life was very much simpler than today. Our customers were fairly well defined, and their needs, at least as well as we knew them, fell into a few large groups. The role of the engineer was primarily about making products available to the mass market, introducing technology and reducing cost along the way.

The market we all enjoyed was stable, with the focus very much on domestic business. It was heavily constrained by regulation and various other barriers to entry.

As a result, we faced the same competitors each year. If we were able to steal a march on them with a particular innovation or new product, we knew that we would have the opportunity for a reasonable period before they were able to catch up. Products continued from year to year, with relatively minor changes in functionality, performance or price. In BT for example, the main product offered was the telephone call. The pace of innovation can be judged by the fact that the standard "Strowger" electro-mechanical exchanges, first introduced in 1912, were still used by 98% of our customers 60 years later. Today of course, the "Strowger" is no more, and our telephone exchanges are essentially computers, with the software updated a number of times each year to support the ever-expanding range of features and services our customers expect.

The stability and consistency of business life really reflected the world around us. Most people fitted, or thought they fitted, into broad social groupings which in many instances laid down the pattern of their lives. Society at large appeared to be stable with common needs and an underlying sense of continuity.

But of course continuity implies the absence of change, and the comfort of predictability, consistency and simplicity. These comforts need to be set against what was missing: flexibility, choice, innovation and growth. We now see that these are the factors that are shaping the present and more and more are going to shape our future.

ANTICIPATING THE FUTURE

Let me look then in a little more detail at the world that is to be, at the changes that are going to shape our lives as we emerge into the age of the Information Society.

On a world scale, there seems to be little reason to doubt that the economic and market trends of the past few years are set to continue for the foreseeable future. The developed economies of the West can expect a relatively stable economic and political environment with continuing growth and relatively low inflation. Growth rates in the developing economies of the East may slow but will continue to outstrip those of the West and will be increasingly important markets. All in all the globalisation of world markets, one of the biggest changes in the past decade, is going to continue as markets open up and as increasing trade links economies closer together.

We can also anticipate some new challenges when we look at social trends. One of these is the increasingly aged population, especially in Europe & Japan. In 25 years time the over-65s will form 30% of the adult population in the UK. In Japan that figure will be close to 45%. This is going to create new demands for our society, particularly with regards to health care for example, while at the same time it is going to reduce the size of the working population that is able to support them. The wider implications of this stark demographic change cannot be ignored by any sector of the community.

At a more personal level, a trend that will accelerate is the blurring of the division between home and work. Partly as a result, discretionary free time will continue to decrease for those in full time work. People will be more demanding about making the best of the time they have, about having individual control of their leisure activities.

The nature of work itself will continue to change, with higher mobility, more shared, contract and part-time working and greater demand for teleworking. The skills required will change, with greater demand for professional, technical and service skills - all areas that are facing shortages today. In fact, demand for IT engineering and systems analysis skills is forecast to double each year for the next decade at least. Re-training and re-skilling are becoming part of every day life.

THE ROLE OF TECHNOLOGY IN DRIVING CHANGE

There are a number of factors that are driving these changes, but perhaps the single most influential factor is the remarkable rate of progress we are witnessing in technology, and particularly in information technology.

People have used technology of one kind or another to manage information throughout history, but IT as we know it today is a very recent phenomenon, reaching back even on the most generous definitions over less than a third of this Institution's history.

In that short time however IT has had an enormous impact on our world. In less than 50 years we have moved from a few massive computers in the world's top laboratories to the point were information technology is all pervasive. We have computers in our homes, in our cars, in our watches and even in our credit cards.

Progress in virtually all the key technologies continues and in some areas is accelerating, giving ever better performance at lower cost. Much of this is captured in the famous "Moore's law", which was defined by Gordon Moore, one of Intel's founders, in 1965. This law says that miniaturisation of electronics doubles every 18 months or so. It has held good for the past quarter of a century, from about 2,000 transistors in the first microprocessor in 1971 through to more than 5 million in the latest Pentium processor. There is every reason to believe that this pace will continue for another decade at least. Combine this with continuing deregulation and the real prices of IT goods and services will continue to fall for the foreseeable future. We can also expect the lifetimes of technology, and hence the lifetimes of products and markets, to continue shrinking.

Another important trend is the continuing spread of digitalisation, as this brings the advantages of digital technology to more and more areas of our lives. We are just starting to reach the point where all forms of communication can be supported digitally in the mass market. Speech was the first to feel the benefits with the quality and cost reductions of digital telephony having been available since the mid 1980s. Digital music followed in the late 1980s with CDs taking over the consumer market - no doubt digital broadcasting, rolling out now, will have a similar impact. Digital writing, in the form of e-mail, is starting to dominate the business world and is even beginning to make a significant impact on the consumer market. And finally we have digital pictures with the first mass-market digital cameras, and the digitised special effects in TV and film-making that we increasingly take for granted.

A consequence of this massive increase in digital content is that demand for high-speed communications is rising rapidly, as having created content you need to get it to your audience, the paying customers. Manageable, cost effective bandwidth is becoming a key requirement. Competition in telecommunications markets is bringing affordable bandwidth into our core networks, and increasingly is pervading our access networks. Here again technology is a critical enabler, as the use of compression technology increases the capacity of our existing copper networks. For example, a 20 Mbyte video clip that takes an hour and a half to download over one of today's top-of-the-range analogue modems will take less than 5 seconds to download over a cable modem or digital satellite service. New protocols are already enabling reasonable quality web broadcasting. Real-time, interactive, high-quality multi-media is almost here!

In mobile telecommunications, people are getting used to being able to get in touch from wherever they are. Initially it was predicted that customers would abandon fixed networks to join mobile ones. In practice however, the shape of the market is changing with better technology and more experienced customers, and what we are now seeing is a convergence of fixed and mobile markets. We are also beginning to see a rise in the use of data in mobile communications, rather than just voice. The widespread availability of the digital GSM standard has led to cheap links for people on the move to use via their laptops, while back in the office we can see sales of wireless telephone and local area network systems growing at 30% per year between now and the millennium.

Satellite is another area where technology is driving change in a major way - a whole host of new possibilities are emerging.

At one end we have the relatively low-bandwidth voice and data services that will be accessible from any point on the earth's surface through strings of low-earth orbit satellites currently being developed by several major global consortia. Then - we have digital satellite broadcasting, bringing hundreds of channels to the home at low cost. And finally we have the prospect of using some of these high bandwidth satellite channels to provide very high speed connections, for example from the Internet, to individual homes or businesses.

THE INTERNET - A MODEL FOR THE FUTURE?

While looking at technology, it is worth reflecting on the Internet in it's own right, because of the impact it has had in such a short space of time. The impact is obvious to all of us, and many would say that the Internet (and intranets, the in-house equivalents used by businesses today) are the foundation of a truly global information infrastructure. Many believe that the Information Superhighway, rather than arriving with a big bang, is sneaking up on us. An interesting model for future rollout of other technologies perhaps?

The success of the Internet, and the World Wide Web in particular, has made everyone in business reconsider their plans. It is driven by truly open international standards that are not owned by any particular commercial interest. It combines a relatively strong underlying technical design with a relatively user-friendly graphical front end. It gives low cost universal connectivity to all users and has proved to be a superb platform for harnessing innovation. A platform that less than 5 years ago was used mainly to link text documents for academic and military users has today become a vehicle for low cost communication of sound, graphics and film. The Internet of course still has weaknesses that are being addressed. There is no way of guaranteeing quality of service at present, capacity management is poor by network standards and security issues remain to be resolved. But the commitment of BT and others to Internet Protocol technology, and the investments we are making as a result, will make the Internet an increasingly secure way to transact business around the world.

The importance of usability can't be overstated. Looking back to 1878, Alexander Graham Bell wrote

"the great advantage it possesses over every other form of electrical apparatus consists in the fact that it requires no skill to operate the instrument".

The "it" he refers to was of course the telephone, which unlike the telegraph that preceded it did not require that the users learn complex codes before they could communicate. He could just as well have been referring to the World Wide Web however. We find that young children today can navigate around the world using the simple "point and click" interface that has replaced the complex strings of hieroglyphics that were required even 5 years ago. It really is becoming literally "child's play", - as those of us with young families will I'm sure recognise!

The numbers are hard to pin down, but something like 30 million people were using World Wide Web browsers within 3 years of their invention. Just think, it took the television industry almost 30 years to reach the equivalent level of penetration. Indeed, last year for the first time, the US saw the number of PCs shipped overtake the number of new TVs sold. And we are only at the dawn of this new age.

THE IMPLICATIONS OF CHANGE

For the world of business, as I have already indicated, the overall message is that competition is increasing. Globalisation, as many of you in the audience will have found, is a double-edged sword. It opens new markets in other parts of the world, but also means overseas competitors are free to enter your traditional markets. Convergence and deregulation mean that boundaries between markets are blurring, demanding continuous segmentation and re-segmentation.

To give one example, - entertainment and education, were quite separate in the more formal world of the past. Today we expect to be entertained as we learn. In the financial world, banks, insurers and building societies were very separate in the past. Today, deregulation and technology allows each to offer the services of the other.

Technology, then, continues to lower the cost of entry to markets, and what is more, leading edge technology quickly becomes available to all, making it much harder to maintain a pure technological edge. If you want to set up a telephone company today for example, you don't have to build a network across the country or develop a mainframe billing system. Deregulation allows you to lease resources from other players and concentrate on adding your own winning edge. Today's innovation becomes tomorrow's commodity. The only way to win is to keep ahead of the cycle, to keep innovating.

The pressure of competition in open, technology-related markets really can't be overstated - it changes the rules of the game beyond all recognition.

Let's look here in the UK for example, where competition has reduced telecommunications prices by more than 50% in real terms over the past 12 years. This reduction refers just to the base prices - most of BT's business customers and almost half of our residential customers make further savings through our various discount schemes. The key point to note is that technology per se will not help you to win. It is the human elements such as how you manage innovation, or how flexible your people are in the face of change that will separate winners and losers in this contest.

These forces are driving businesses to concentrate on what they do best, to use the skills and assets they have to differentiate themselves from others. This refocusing creates new opportunities as companies outsource activities that were once considered key internal functions. It also changes the nature of business relationships, increasing interdependence, as businesses rely on strategic suppliers to provide mission critical services and move from providing total solutions on their own - to joint packages with partners and alliances.

For some businesses, the process of what has become called "disintermediation", the removal of middle men, is a significant threat. If consumers are able to scan the Internet for details of holidays, homes or insurance schemes will they need a separate travel agent, estate agent or broker to do it for them? What they may need instead are people who can create the electronic tools and guides that they can use themselves - so again, new opportunities are created.

For other businesses, there is the recognition that you cannot do everything, everywhere, operating on your own. This idea is at the heart of BT's strategy as we respond to the challenges of globalisation. Over the past four or five years we have built up a number of very close alliances with selected partners around the world. These allow us to offer a range of services that are consistent around the world, but that are delivered by a local company that understands the needs of that individual market. Our proposed merger with MCI, one of the US's most innovative and highly respected operators, takes us on to the next stage of that strategy.

Looking beyond the world of business, for society as a whole we can see a wide range of challenges but also a lot of possibilities.

Health is already very high on the public agenda, certainly here in the UK, and will become more so in the coming decades as a result of the demographic changes I mentioned earlier. This, coupled with rising consumerism and limited resources, are having the effect of driving demand up - faster than supply. It is vital therefore that we make use of scarce resources, and with the help of IT we are seeing a growing number of applications in the field of tele-medicine. BT's *ISDN* services for example are being used already for remote foetal scanning, where prospective parents in one part of the UK are "examined" in their local surgery by specialist consultants working in the London teaching hospitals.

Education is another area of hot debate. Now more than ever our children need to be equipped with the skills to succeed in a complex and ever changing world. And it is not just children of course, as continual change means that learning is a life-long activity, rather than something confined just to our youth. Again, technology is bringing about hosts of new possibilities as we enter the information age. We already have unprecedented access to (relatively) affordable information. The kind of comprehensive encyclopaedia that would have cost hundreds or even thousands of pounds just a few years ago is available for less than fifty pounds to any household or school with a multimedia PC. Of course this is forcing restructuring in the publishing industry, as some of the traditional encyclopaedia publishers are seeing their revenues falling by 30-40% a year as a result. And this is only the start - moving beyond CD-ROM, access to the unparalleled resources of the Internet shows massive potential.

To quote one young user of BT's *"CampusWorld"* service

"The town library is sometimes closed. At the school library we simply don't have the resources. But CampusWorld is the door to the Internet, the largest library in the world".

In the field of leisure we can expect to see an ever wider range of options, with greater control for the individual. BT have conducted one of the world's most advanced market trials of interactive technology, and the results showed enormous potential for multimedia services that can be customised and delivered "on demand". Most of the UK's national newspapers are available on-line today, and several of them allow you to create a personal profile that gathers the news most likely to be of interest to you.

Of course, all this requires that people are happy and able to use the technology that is available. With the pace of change and innovation that we are witnessing this is a significant challenge. People, especially young people, are adopting new technologies at a faster pace, but many are finding it hard to keep up. It is often a case of building one technology on top of another. We have seen this at BT with some of our network services. The success of our "1471" service, for example, that allows callers to know who has called them paved the way for new services such as Callminder, the "answering machine in the network" and "Ring Back When Free". It is all a question of helping people to see the benefits that are available, without worrying them about the technology.

THE ROLE OF THE ENGINEER

So perhaps now is the time to revisit the role of engineering. Our role is critical, but it is also one that needs to evolve to reflect the changing world around us. Throughout history the engineering profession has been at the forefront of change, but to maintain that position we need to focus on a few key areas

- *Harnessing innovation* is a core skill, doing things in radically new ways. That applies to the way each of us works as well as the things we produce.

- *Getting to market* is critical. A six month delay in time-to-market nowadays can make the difference between success and failure. We need greater pragmatism in building what we need to get started, rather than the fully-engineered solution.

- *Commercial decision making* is essential. Our decision making must take into account the value-add of our work, the structure of our costs and the profitability of our products.

- *Exploiting standards* is another core skill. In a future where choice, interworking and flexibility are the keywords, the ability to influence and exploit open standards is critical.

It may seem a little strange that in an increasingly technological world we are placing so much emphasis on our engineers developing their non-technical skills. Look at any large engineering organisation and the emphasis is on better teamwork, closer working with customers and suppliers, streamlined development and delivery processes, greater focus on cost. Indeed it may seem that we are being asked to hide our light under a bushel, to be almost embarrassed about the technology we understand so well.

But it is because technology (and information technology in particular) is so pervasive nowadays, because it is so critical to the way we live our lives, and because so many people are overwhelmed by the profusion of choices that face them that the engineers role is so important. It is really that of guide and helper, delivering services that are friendly to the average user, that allow them to navigate the complexity underneath, and that take them reassuringly to their goal.

- END -

uilding technology

ALAN COCKSHAW FEng
airman, AMEC plc, Manchester, UK

The 150th anniversary of the Institution is a time to recognise the many fine achievements of the past but more importantly to explore how, in a more effective and co-ordinated way, we can foster the further development of technical excellence across the broader industry we all serve.

The IMechE has common ground with the range of professional institutions which represent the engineering and construction industry, both at home and overseas. Where such common goals exist we should look to reap the benefits of closer association and avoid dilution of effort. We can achieve so much more by working together than we have ever done in our separate endeavours.

Any attempt to consider future changes in the way buildings are conceived, designed, and procured must include the examination of three key areas. Firstly the process by which they are completed; secondly potential changes in architectural form and structure and thirdly the changing mechanical and electrical engineering content of all of them coupled with rapidly evolving safety and environmental control systems.

If we examine the process in the United Kingdom, most of the professionals involved accept the need for radical change. Overall, most would also agree that in recent years the process of delivering new structures has been to say the least, inefficient. In simple terms the industry has not been delivering value for money.

The much-discussed Latham Report has provided a widely accepted focus on cost reduction and set apparently ambitious but most certainly realistic targets for the industry to achieve. The challenge of the Latham initiative has been taken up by many of the key organizations in the industry but the quality of the response so far has been extremely variable. Comparisons between the United Kingdom and, for example, the United States and South East Asia make it clear that the potential improvements suggested in the Latham Report are desirable and achievable. They should be implemented with urgency if companies in the industry are to compete with other nations not only in the United Kingdom but in the international market place as a whole.

In recent years Britain has produced some of the finest architects in the world. Practices such as those of Foster, Rogers, Grimshaw and Farrell have the kind of vision that clients from all over the world want to use. The fact that these architects are in such demand around the world is testimony to their quality. This is equally true of some architects in the United States, South East Asia and, of course in mainland Europe too. In Britain today we have not only excellence in our architects but also in our structural, mechanical, and electrical engineers who have immediate access to the increasingly sophisticated technology and systems that make world-class buildings function the way they are intended.

We already have, therefore, the people and the tools to match our future needs. The immediate challenge is to fully harness the technology and talent already available to realise projects far more efficiently than we have in the past. So what is wrong with the process ?

Until quite recently the development of buildings in the United Kingdom has been perhaps noted for the separatism of all those involved in the process. Firstly, in design, between the architects, the structural engineers and the mechanical and electrical engineering specialists. Secondly, the separatism between the design and construction processes and, thirdly, the historical inability to adequately consider a lifetime cost approach to new projects.

The Offshore Oil and Gas Industry, for example, which in the past has suffered from many of the weaknesses of the general construction industry, has been revolutionised by the recognition that the separatism of the participants in their contracts has resulted in an escalation in the cost and time of all of them. The key companies in this sector are, in the main, global players and their procurement techniques have been developed from their experiences in many parts of the world. By and large, they have increasingly rejected the old separatist approach in favour of partnerships and alliances which have now become the order of the day with major improvements commonly achieved in value for money in their capital projects.

By comparison, those who procure major capital building projects vary considerably. The larger, more experienced developers and building owners have long recognised the need for change, and are constantly seeking improvements to the process. However a significant percentage of all the buildings procured are procured by clients who perhaps only do so once or twice in a lifetime. In such cases, the understanding of the benefits to be gained by adopting a more modern approach to the development of their project is difficult for them to recognise. After all, every building is unique within itself. It is extremely rare to build two of anything.

What every client needs in today's market is greater certainty in the delivery of their projects in time, cost and quality. Clearly these objectives have to be met whilst sustaining the creative vision developed and agreed between the client and the chosen design team.

The integration of all of the professionals involved in the process of delivering new buildings into a single-point of responsibility, maximises productivity and improves communication immensely. Whether that approach is called design and build, guaranteed maximum price, partnering or other similar concepts, these approaches consistently provide a more effective process.

In such contracts, all the parties have a common goal and providing that the definition of the project is properly detailed at the outset, the project will be executed properly and efficiently. A multi-layered approach with each party shifting risk and responsibility is not only costly but time-consuming. Regrettably, it is extremely beneficial to the legal profession but not to the procurers and executors of the projects themselves.

The whole process has been very inefficient and the future will most certainly require far more effective integration of all the skills of the parties to the development of the buildings of to-morrow.

A fundamental area of improvement lies in the application of information technology which has not yet had the impact in the industry that is capable of being achieved. For example, modern, sophisticated computer-aided design has too often been used only as a drawing tool. The improved efficiency and speed of production has brought many benefits but the potential for integrating computer-aided design across all the architectural and engineering disciplines has not yet been fully realised.

Although many current projects are designed using 3D CAD systems, many are not. The use of virtual reality programmes allowing clients, their architects and all of those involved in the process of creating the building to understand and interrogate the way in which the building will be used when completed, is a particular area where substantial improvements can be gained. The minimisation of changes and additional works is clearly a key objective.

Again the Oil and Gas Industry has established these techniques very effectively, contributing hugely to the identification of problems and the promotion of effective solutions. Procurement systems, quality assurance, shop drawings, measurement and costing processes automatically follow from the integration of all these systems. With all parties to the contract "on-line" so to speak, communications have improved dramatically and projects are executed quicker and far more cost-effectively. The technology exists to achieve all of these objectives in the broader construction industry, where previously the separatism of the parties has prevented effective implementation.

Whilst construction has certainly lagged behind other industries in the co-ordination of its key elements, the changes to the process involve everyone's acceptance of the need to spend far more time in the beginning in interrogating the design processes to avoid unnecessary additions and changes in the scope of work to be executed.

The second key area we must examine relates to form and structure. In this regard, some leading providers of major modern buildings in the United States would suggest that building technology has perhaps changed very little in the last 20 to 30 years. Many structures, of course, have become lighter, using higher strength materials. New facades have developed, often with more glass and geo-technical developments have facilitated newer, more cost-effective and time-effective solutions for foundations.

Recent years have seen some exciting developments in architectural thinking and we can expect even more stimulating challenges in the years ahead. However the link between architects and structural engineers, in particular, is now much closer than it was some years ago and the development of new visions are now almost always accompanied by effective conceptual structural solutions.

Continued acceleration in the use of three-dimensional CAD systems is encouraging new approaches to space planning. These are essential tools as the historical ratios of space requirements in new buildings change especially in high cost city centre areas.

Overall, although the technology required for what is commonly referred to in the United States as the "shell and core" of our buildings will continue to improve. The changes we might anticipate are perhaps rather more incremental than fundamental. Clearly the more sophisticated solutions require greater use of modern information technology, to bring together all of the technical disciplines more effectively.

Of the three key areas, the one which will present the greatest challenges in the future is what is inside our buildings. When we look at the changes over the last 20 years, the basic mechanical and electrical engineering content of buildings has certainly become increasingly sophisticated. However, as we look forward, it is obvious that modern technology offers enormous potential for future development and that application has fallen far behind the opportunities available.

A modern airport terminal building, for example, is a particularly good illustration of the breadth and depth of change which is now being rapidly introduced. Security systems, baggage handling systems, information systems, people movement, freight movement, safety and security must all be accommodated whilst maintaining a smooth flow of large numbers of people in a user-friendly environment. The design, implementation and continuously developing needs of a modern airport presents a particularly good illustration of the way in which architects and engineers are beginning to use the best available technology to satisfy the rapidly changing needs of an industry where the growth in traffic and the needs of the airlines and their passengers are constantly developing.

Tomorrow's buildings must be able to embrace all of the tools that companies need to meet their specific challenges in the global market. Yet they must be user-friendly. Tenants and owners are looking for buildings which provide a more secure environment. One which will allow them to communicate effectively with their other offices, their customers and their suppliers around the world, literally instantaneously. They are looking for energy-efficient offices and those with ever more sophisticated environmental controls with potential problems, such as the "sick building" syndrome, designed out.

Greater attention is now being paid to the lifetime costing of new developments; to the examination of operating costs and dedication to their continuous reduction. The concept of fully intelligent buildings is not new but they have been very expensive to provide. The challenge now is to provide those benefits in a more cost-effective way.

Sophisticated computerised building management systems are now readily available to maximise the efficiency of the electrical and mechanical utilities in a building as well as to provide increased security and protection. Remote monitoring of the key systems is already with us but these will improve substantially as the technology is perfected and the cost of implementation becomes more competitive.

Too often the implementation of these systems follows completion of the building rather than being an essential design ingredient at concept stage.

It is often said that, within the United Kingdom, buildings are over-designed and over-specified in comparison with other parts of the world and in many cases unnecessarily so. More research work could beneficially be carried out to examine why, for example, buildings in New York are so relatively cheap compared with buildings in the United Kingdom for equivalent quality.

Our buildings are essentially designed to meet existing codes, which in truth often lag behind the technology that is available. This represents considerable challenges for mechanical and electrical engineering innovation, which again will be significantly enhanced by closer association with the architects and structural engineers involved in the development of the building form.

These three key ingredients of process, structure and contents are clearly not separate issues; they are all closely integrated but improvements in the process are central to the continuous improvement of everything else. We must use the available technology much more effectively than we have in the past. We must have an integrated approach to design with maximum utilization and integration of CAD systems, with the same data base flowing through from design, into procurement and execution. Planning systems, quality systems, cost systems must all flow freely between those involved in the process. A far more integrated approach with everyone working together towards the common objective of delivering the vision in the most cost-effective, time-effective way.

Commitment to these concepts by all those party to the contracts will significantly accelerate delivery, reduce costs and create the climate within which clients achieve better value for money, whilst allowing all of those involved in the process to get a better return for their efforts in completing the project.

This philosophy, of course, requires far greater emphasis on the detailed front-end engineering and planning in order to reduce the incidence of additional and varied works which so often lead to the development of the kind of adversarial relationships so prevalent in the past.

More emphasis must be placed on guaranteeing the completion date and the completion cost and much less on the start date. Much more time spent in design and planning and less in dispute argument and resolution.

Tomorrow's buildings will need to be more flexible, more responsive to the global markets in which the occupiers are engaged. They must be more cost-effective to deliver and more economic to run. As the industry begins to apply established technology in a more effective way, fundamental improvements can and will be made in both capital expenditure and operational costs. Thereafter progress will be more incremental than fundamental.

Finally, a particular word about the United Kingdom, where compared with other parts of the world, we have a Town Planning system which is almost entirely reactive to the individual proposals of particular developers, which requires either agreement, rejection or modification. It would be infinitely better if we created a more proactive approach to the overall built environment to the greater benefit of all those who invest within it.

e engineering challenge

DAVID DAVIES CBE, FEng, FRS
sident, The Royal Academy of Engineering, UK

The 150th Anniversary of the Institution of Mechanical Engineers clearly provides an opportunity to stand back and review the enormous contribution engineering has brought to the world in order to learn from history and before we develop a vision for engineering in the future. The programme put together by the Institution for this Conference concentrates on key areas in which engineering is having substantial impact on world development. The programme for the first day comprises a group of overview papers that effectively set the scene for the whole conference.

Despite the fact that these papers cover very different aspects of engineering, there are many interesting common themes that emerge. For example, in the transportation field, there are difficult areas of compromise to be struck between passenger convenience, (which determines usage) energy utilisation, environmental impact and of course passenger and public affordability. This is also an area where the scale of public funding for major infrastructure is of critical importance.

In the manufacturing area we see large structural changes in the industry on an international scale. This is driven by new manufacturing practices based upon IT and often coupled to new materials. Here again we find environmental issues can have a major impact - particularly in terms of constraining legislation. In many instances the most severe constraint on future plans is a lack of adequately trained manpower for the changing industrial tasks. The rapidly changing technological scene emphasises the need for continuous updating of education and skills.

The international dimension also emerges as a particularly important factor in the papers. Although mobility of labour is still relatively limited, mobility of work including that on an international scale, has had a substantial impact on employment patterns. This is driven by international markets, despite the attempts by many governments to construct local regulations for national benefit. The engineering knowledge base is, of course, wholly international and for this reason educational policies of governments can often have more impact on job creation than restrictive legislations.

One of the interesting debates which has recently taken place within The Royal Academy of Engineering concerns the boundaries of engineering. As an interdisciplinary body The Academy has asked how such boundaries are defined and whether they are changing. I have certainly concluded that sometimes we have been too slow to recognise and capitalise on new developments in science. The evolution of information science into engineering is a particularly strong example of this phenomenon.

At present we see some of the most rapid and far-reaching scientific developments occurring in the biological area. It is interesting to ask whether this will lead to new biological technologies in the future. Today's paper on Engineering and Health in Medicine is a very appropriate contribution to that debate. Not only do we see the application of intelligent control systems and nano-technology to surgery and artificial limbs but the whole field of biotechnology and genetics open a new range of opportunities for engineering applications to assist mankind in a very direct and human way. The examples quoted in the paper for new materials in the form of tissue engineering and indeed for genetic engineering are extremely interesting and thought provoking.

Engineering is about "causing things to happen". It is often on a large scale - for example by providing clean water, power or communications in part of the developing world. It can also be on a smaller scale such as the provision of an artificial limb which can have important personal benefits to individual lives. Engineering is often about advanced specialist technology, but it is also about the interdisciplinary application of such technology more broadly within society.

It is interesting to note that the 150 years of the Institution's existence has witnessed worldwide effects of the Industrial Revolution. This has involved new engineering developments in the form of mechanical and electrical power sources which have transformed both transport and manufacturing. This Revolution came about as a result of enhancing and extending the physical capabilities of human beings. It produced profound changes in society. Many new industries and a complete transport infrastructure was created. It changed where people lived and the nature of their work and leisure. It started primarily in the UK and other parts of Europe but it has been world wide in its impact although there are still many areas which have so far been only lightly touched.

We are now in the process of an Information Revolution. This is able to apply intelligence to many machine processes and thereby extend and enhance the intellectual capability of human beings. As yet, the scale and overall consequences of such a Revolution is difficult to visualise. However, the impact on employment and on society is as least as far-reaching as that of the Industrial Revolution.

What is more important is that current changes are now happening at a far faster pace and in a highly interactive industrial environment. Indeed the international dimension of engineering is a factor which has dramatically affected all the topics in this event. Manufacturing products, investment and indeed knowledge move internationally in response to world markets. We must be careful to appreciate the effects of such factors, particularly when trying to introduce changes in one country.

There is no doubt that over the past decade society has developed an extremely strong interest in the environment and quality of life. Unfortunately this has sometimes been expressed as a criticism of all science and engineering as a result of past examples of unattractive environmental outcomes. However, we must also welcome this interest since engineering offers excellent opportunities to improve the environment and the quality of life. It is the engineers role to offer the full power of engineering development to achieve such desirable outcomes.

Engineers have to make it clear that the way forward involves priorities and choices. Education is of critical importance here in making young people aware of improving the quality of life as well as the opportunities for devaluing it. Again the international dimension is paramount. Sustainable development is an issue which has to be taken forward on a global scale. It is an issue requiring partnership between the developed and the developing sectors of the world. Innovative manufacturing ideas may be introduced in one country but exploited in another. We have learned many lessons from the Industrial Revolution but we will all have to learn much faster to cope with the current rate of change as new engineering technologies are applied internationally.

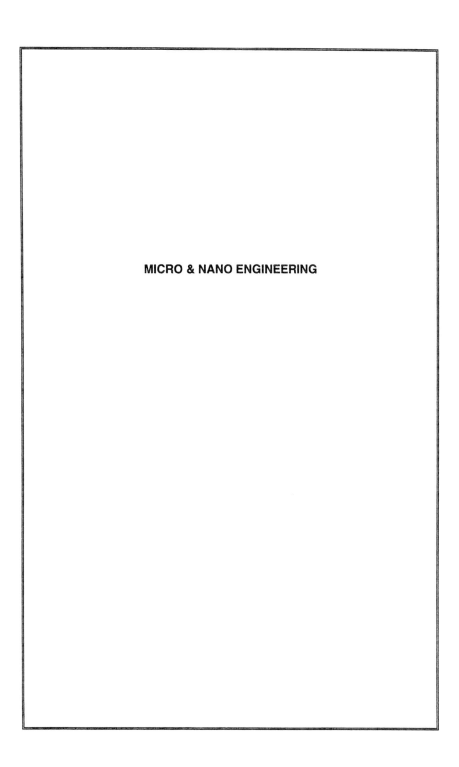

MICRO & NANO ENGINEERING

om micro to nano technology – an international
rspective

ritus Professor P McKEOWN OBE, Hon DSc, MSc, FIEE, CharterFSM, FIMechE, FEng
field University, and Precision Engineering Consultant at Pat McKeown & Associates, Bedford, UK

1. FOREWORD

This is the "150th" Symposium session which addresses advanced manufacturing technology, a traditionally strong interest of the Institution. In this, the third Industrial Revolution, manufacturing technology which has contributed so extensively to the standard of living we enjoy today, is now set to take us into what has been called the 'Nanometre Age' (1). During the last 150 years, manufacturing technology has developed almost beyond recognition in terms of processes, machinery, organisation, control and management. The main vehicle for this dramatic change has been **automation** under the driving force of **wealth creation at lower overall cost**. It is most appropriate that on the 150th anniversary of our Institution we should address the next major innovation in manufacturing - an entirely new approach, a new way of thinking, that is rapidly emerging; this is **micro- and nanotechnology**.

Several existing products will be replaced by highly miniaturised versions - and many new products will appear, made possible by micro and nano processing technologies that over the next 10 - 20 years will progressively replace in several areas, current manufacturing processes and factories.

Micro and nano engineering are, in a sense, subsets of **precision engineering**, a development that has accelerated over the last 30 years in terms of research, development and application to product innovation. It has been driven by demands for **higher performance, higher reliability, longer life** and **miniaturisation**.

Miniaturisation in many areas offers the prospects of products that are "**smaller, faster and cheaper**" which is certainly true of integrated electronic micro circuits, VLSI and ULSI. It is also true for **microsystems technology (MST)** products that would be impossible to make by **today's macro-engineering** processes. Many MST products frequently also referred to as micro electro mechanical systems (MEMS), especially in the USA, are now in widespread use.

This session describes the state-of-the-art in ultra precision manufacture of micro and nano technology artefacts in the fields of engineering and bio-medicine. Both are examples of improving quality of life through technology.

2.   MANUFACTURING

The main wealth-creating activity of industrialised nations is manufacturing. Manufacturing dominates world trade. Since 1950 world trade in manufacturing has grown tenfold and the fundamentally tradeable nature of manufactured goods has created a dynamic engine for growth in the most successful manufacturing economies. In the UK, manufacturing employs more than 5 million people and produces 21 per cent of GDP (USA 20 per cent; Japan 28.2 per cent; Germany 32 per cent); it is the largest single element of the UK's economy. Manufactured goods represent 62 per cent of all exports (greater than the exports of banking, insurance and oil combined). A large proportion of the activities of the service industries depends heavily on manufacturing, e.g. banking, catering, retailing and distribution and telecommunications. A further 5 million people in the UK have jobs indirectly dependent on manufacturing.

When asked what has contributed most to the dramatic improvements in standards of living (if not quality of life), which most of the industrialised countries have enjoyed over the last 40 years, most people reply that it is primarily **the new products of technology**. Examples given include:

- domestic appliances such as refrigerators, microwave cookers, automatic washers, television, VCRs, hi-fi stereo systems, convenience foods;
- healthcare, pharmaceuticals etc;
- cars, high speed trains, aircraft;
- computers, telecommunications, satellite TV, fax and copier systems.

It is not widely appreciated that it is development in the **methods of producing these products**, i.e. **advances in manufacturing technology (AMT)** and its management - not just the products themselves - that have had such dramatic effects on our culture in terms of increased living standards. Woven garments were worn in antiquity; it was the mechanical loom that made textiles into consumer products. In short, machines and mechanisation started a revolution leading to mass production and the associated wealth creation and prosperity.

The Second Industrial Revolution started with Henry Ford's production lines and thence the evolution of advanced manufacturing technology (**AMT**) into basic automation. We are now in the Third Industrial Revolution where overall corporate system performance optimisation is the order of the day. Innovation in products, manufacturing processes, machine systems, coupled with world class efficiency, is key to future success. In fact, the future will be based on high technology and this needs an innovative, efficient manufacturing base.

A major innovative development within AMT and thus addressed by this session is **microengineering and nanotechnology**.

3. HIGH PRECISON MANUFACTURING

Manufacturing with higher precision has accelerated over the last 25 years in terms of research, development, and application to product innovation. It has been driven by demands for **much higher performance of products, higher reliability, longer life and miniaturisation**. This development is widely known as **precision engineering** and today is generally understood as manufacturing to tolerances smaller than one part in 10^4 or perhaps 1 part in 10^5 (1). This is the 'umbrella' term under which there are the important major sub-sets of **microengineering** and **nanotechnology**.

Microengineering, also frequently referred to as micro systems technology (MST), is where the physical dimensions of the component or its features are small, namely in the order of 1 micrometre (1 μm). This is covered comprehensively by Dr W Ehrfeld and Professor R A Lawes in this session.

The historical roots of precision engineering could be said to be horology, the development of chronometers and watches together with optics, particularly the manufacture of mirrors and lenses for telescopes and microscopes. Major contributions were made to the development of high precision machine tools and instruments in the 1800s by ruling engines for the manufacture of scales, reticules and, later, spectrographic diffraction gratings. Today, ultra precision machine tools under computer control using single point diamond tools or multi point diamond grinding wheels can position the tool relative to the workpiece with a positioning accuracy in the order of 1 nanometre (1nm). However, it will be noted in Fig. 1 achievable "machining" accuracy includes the use of not only cutting tools and abrasive techniques but also energy beam processes such as ion beam and electron beam machining, plus scanning probe systems for surface measurement and molecular manipulation (pick-and-place).

So ultra precision manufacture has progressed through micrometre (μm) accuracy capability to enter the nano scale regime.

The term '**nanotechnology**' was first introduced by Professor Norio Taniguchi formerly of Tokyo Science University, in 1974 at the ICPE Conference Tokyo (2) (3) (4). He used the word to describe ultra fine machining - the processing of a material to nano scale precision, work that he started in 1940 by studying the mechanisms of machining of hard and brittle materials such as quartz crystals, silicon and alumina ceramics, primarily by ultra-sonic machining. Subsequently he has been in the forefront of research in using energy beam processes (EBM, FIB, RIE etc.) which he sometimes refers to as 'atomic bit' processing for fabrication of such materials to nanometric accuracies. However, it can be argued that the **concept of nanotechnology** was first enunciated by the American physicist Dr Richard Feynman (later in 1965, a Nobel Laureate) in his visionary lecture given in December 1959 at the annual meeting of the American Physical Society at the California Institute of Technology, Pasadena, California. He entitled his talk "There's plenty of room at the bottom", (5). At the outset he asked, "Why cannot we write the entire 24 volumes of the Encyclopaedia Britannica on the head of pin?". He pointed out that if you magnify the head of the pin by 25 thousand times, the area would then be equal to the area of all the pages of the Encyclopaedia Britannica, and went on to point out that all that was necessary was to reduce the size of all the writing in the encyclopaedia by 25 thousand times. He argued that the scanning electron microscope could be improved in resolution and stability to be able to "see" atoms and went on to predict the ability to arrange atoms the way we want them, within the bounds of chemical stability, to build tiny structures leading to molecular or atomic synthesis of materials. On all counts his predictions have been remarkably accurate, (for example, see Fig. 5). He did not use the term nanotechnology as such, but accurately described its potential for extreme miniaturisation and the self organising and self assembly of molecules, i.e. 'bottom up' nanotechnology possibility of mobilising nano-scale molecular structures to act as machines to guide and activate the synthesis of larger molecules. This is now called "**molecular nanotechnology**" the leading promoter of which is Dr Eric Drexler (6).

ACHIEVABLE 'MACHINING' ACCURACY

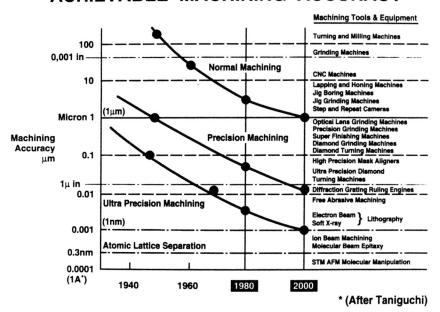

* (After Taniguchi)

Figure 1   The development of achievable 'machining' over the last 60 years

Perhaps the best definition of Drexler's **molecular nanotechnology** is "**the projected ability to use potential control of chemical reactions to building complex materials and devices (including molecular machinery) resulting in precise control of the structure of matter at the molecular level**". Many engineers and scientists remain somewhat sceptical about these ideas for "building with molecules" and referring to Drexler, as a 'futurologist', warn of the dangers of fuelling unrealistic expectations.

A more broadly accepted working definition is "**the study, development and processing of materials, devices and systems in which structure on a dimension of less than 100 nm is essential to obtain the required functional performance**". For the UK's Link Nanotechnology Programme (1988 to 1994), it was succinctly defined as "**design and manufacturing in the tolerance range from 100 nm (0.1 μm) to 0.1 nm**".

Fig. 2 Scanning electric micrograph of 'chip' of electroplated copper cut by a sharp diamond on an ultra precision machine tool. The underformed chip thickness is about 1 nm. The resulting surface smoothness is about 1 nm Ra - a super smooth, very high reflectivity surface is produced directly on aluminium, copper, electroless nickel etc. without the need for subsequent polishing. (Courtesy Professor N. Ikawa, Osaka University).

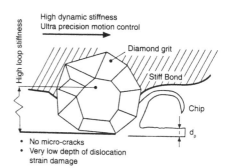

Fig. 3 The mechanism of "ductile" or "shear mode" CNC grinding of brittle materials; depth of cut d_p must be less than the critical transition depth at which brittle fracture occurs. This demands high precision, high stiffness machine tools and achieves very high surface smoothness with virtually zero sub-surface microcracks.

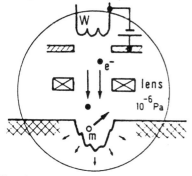

Fig. 4 The principle of electron beam machining (EBM); high energy electrons are focused electromagnetically onto the target in a vacuum; nanometre patterning accuracy can be achieved.

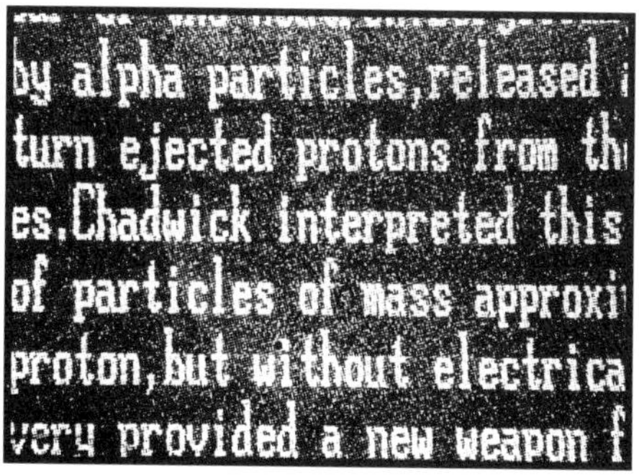

Fig. 5 High speed raster scan EBM "writing the Encyclopaedia Britannica on the head of a pin"; the line width of each letter is two electron beam drilled holes; each hole is 4 nm diameter; substrate is silicon fluoride (Courtesy Professor C Humphreys, Cambridge University.

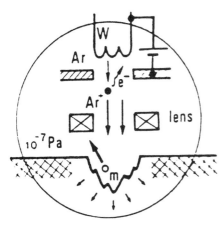

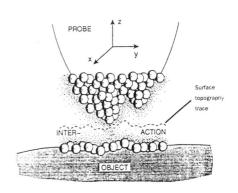

Fig. 6 The principle of ion beam machining; electrically accelerated inert-gas ions such as argon are projected at the target surface in a high vacuum, providing "atomic- bit" machining capability.

Fig. 7 Schematic of a scanning tunnelling microscope probe, showing interaction of the electron clouds of the nearest atoms of the probe and object. The probe is moved in the X, Y directions and servo controlled in the Z direction with picometre resolution - to produce a Y scan trace of the surface atoms (topography) of the object.

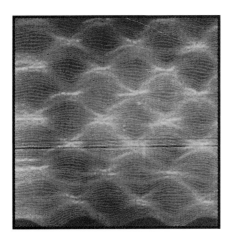

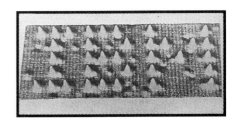

Fig. 8 The screen image of an STM surface scan showing atoms of carbon; this whole image is approx 1 nm x 1nm.

Fig. 9 SEM image of 35 Xenon atoms on nickel substrate - IBM

4. ULTRA PRECISION MACHINING AND MATERIALS PROCESSING

4.1 Nanotechnology processes, metrology, instrumentation and machine systems are required in the manufacture of most, if not all, components, assemblies of components and complete microsystems in the microengineering regime. Nanotechnology brings together engineering, physics, chemistry and biology. It includes material processing through removal, accretion, surface transformation, joining and assembly, right down to identification, manipulation and assembly of individual molecules.

Ultra precision (nanotechnology) "machining" processes include:

- Single point diamond and CBN cutting (Fig. 2);
- (Multi-point) fixed abrasive processes, e.g. diamond and CBN grinding, honing, belt polishing, including "ductile mode" micro-crack-free grinding of glasses and ceramics and other brittle materials (Fig. 3);
- Free abrasive (erosion) processes, e.g. lapping, polishing, float polishing (mechano-chemical, chemico-mechanical processes);
- Chemical (corrosion) processes, e.g. etch-machining (perhaps after photo-and-electro-lithography);
- Biological processes, e.g. chemolithotrophic bacteria processing;
- Energy beam processes (removal, accretion and surface transformation processes), including:
    - photon[x] (laser) beam: micro-'cutting', drilling, transformation-hardening;
    - electron beam[x]: lithography, welding, micro- and nano-drilling, (EBM) (Figs. 4 and 5);
    - electro-discharge[x] (current) micro-machining (EDM);
    - electrochemical (current) machining (ECM);
    - LIGA deep etching using X-rays for lithography; then electro plating and moulding of micro components;
    - inert ion beam[x] machining (erosion) (focused (FIBM) and broad beam with mask) (Fig. 6);
    - reactive ion beam machining (etching) (RIE);
      thin film techniques (Langmuir Bodgett) (LBTFT);
    - molecular beam epitaxy (accretion) (MBE);
    - scanning tip process engineering (STM, AFM) enabling molecular manipulation, assembly and modification (Figs. 7, 8 and 9).

[x] indicates the process is essentially electro-thermal; for "atomic-bit" processing, high density energy in the range of 10^4 to 10^6 Joules/cm^3 is necessary.

4.2 Scanning probe microscopy (SPM)

The Scanning Tunnelling Microscope (STM) is the best known of all the scanning probe microscopes now developed. In the STM, the electron clouds surrounding the nearest atoms of the electrically conducting tip and object interact and a "tunnelling current" flows. This is amplified and used to servo drive a piezo actuator in the Z direction so as to track the object's surface profile as the tip scans over the object in Y. This is the metrology surface topography instrument mode of operation; it can image surface atoms as shown in Fig. 8.

The Atomic Force Microscope (AFM) is a variant of these SPM "local probes" or "**nano fingertips**", unlike the STM, it can measure non-conducting materials by low force physical contact. However, these local probes can also be used as "machining" tools for manipulating atoms in pick-and-place mode i.e. by extraction and deposition of single atoms or clusters of atoms. Probably the most famous example of this technique is the IBM image of 35 Xenon atoms on a nickel substrate achieved by Eigler & Schweizer at IBM Research Division, Almaden, California, in 1990, Fig. 9. The STM experiment has to be done at cryogenic temperatures ($4°K$) in vacuum. Each atom was "dragged" across the clean nickel surface at a speed of 0.4 nm per sec. Although this is an example of 'atomic pick-and-place' it is far from an economic method for molecular manufacture on an industrial scale.

5. ULTRA PRECISION MACHINES

5.1 Ultra precision machine systems fall into three main classifications:
- computer numerical control (CNC) **macro-machines** for measuring, shaping or forming conventional macro-sized component parts today, this can mean working to nano tolerances on macro-components (see 5.2 below);
- instruments for metrological applications to macro and micro components;
- very small 'Feynman' machines ranging in size from a few millimetres down to micrometre dimensions.

In each case there are Eleven Principles and Techniques that must be used in design and build to achieve under computer control a **highly deterministic performance** as a sound basis for nano precision capability (1), (7).

5.2 Single point cutting and fixed abrasive grinding have been dramatically stretched in the last 10 years to genuinely enter the nanotechnology regime. Fig. 2 shows how on an ultra precision diamond turning machine, an undeformed chip thickness of 1 nm is achievable on high purity copper. This has been made possible by the development and combination of diamond cutting tool materials, ultra precision and highly stiff fluid film (hydrostatic) linear and rotary bearings, ultra precision state-space servo drive technology and CNC with multi-axis dynamic error compensation. Together with multi-loop servo temperature control to $0.01°C$ or better, these have been combined to produce machine tools such as the Cranfield Precision Nanocentre, a world state-of-the-art ultra precision machine tool (Figs. 10 and 11).

It was designed to achieve very high dynamic loop-stiffness between tool and workpiece in single point diamond cutting and 'ductile mode' grinding. This is achieved through optimised machine configuration and the use of high stiffness servo drives and hydrostatic bearings throughout. It is a 3-axis machine, two linear and one rotary. **Servo positioning** resolution of the X and Y axes is 1.24 nm; the rotary 'B' axis for "tool normal" single point turning has a positioning resolution of 0.3 arc sec. Temperature control of all elements of the machine within its insulated covers is held to better than 0.01°C. This machine was chosen as the basis for the Ultra Precision Machining Research Facility for the UK's LINK Nanotechnology Project No. 1, "Ultra Precision Machining Research" (1988 to 1993). This machine has produced under CNC, surface finishes superior to anything so far reported elsewhere in single point diamond turning of several materials e.g. germanium to 0.8 nm Ra. On workpieces such as complex space satellite communications optics of about 300 mm diameter, overall profile accuracy in the order of 100 nm can be achieved. This is just one example of nano tolerances being achieved on macro-components.

The direct CNC "ductile mode" grinding of spherical and aspherical optical components to finishes of 3 nm Ra in BK7 glass is another important recent break-through capability of ultra precision machine tools like the Nanocentre.

Fig. 10 The Nanocentre ultra precision diamond turning and grinding machine

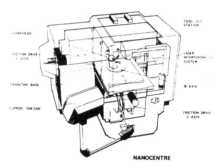

Fig. 11   High Loop stiffness and servo controlled motion to 1.25 nm resolution enables machining of complex, aspheric optics etc. frequently with little or no subsequent polishing etc.

5.3 Undoubtedly the most accurate 3D macro machine in the world today is the **Molecular Measuring Machine** dubbed "M cubed" or just "M3"; it has been researched, designed and developed by Dr Clayton Teague and his team of engineers and physicists at the National Institute of Standards and Technology (NIST), Gaithersburg, Maryland, near Washington DC, USA. This pioneering, world-leading project is aimed at imaging and measuring to nanometre accuracy, the positions of nanometre-size features located anywhere within an area of 50 mm x 50 mm and up to 100 μm high. This nanometrology machine, Fig. 12, is designed to help meet the needs of industry working on next generation nano devices, particularly nano-electronics and nano-ULSI (Ultra Large Scale Integration) circuits. In other words, this machine and others that follow, likely to be built on the same fundamental principles, will provide the metrological standards by which to control the production of ULSI circuits with 100 nm line widths and less, envisaged for widespread use within the next 20 years, e.g. line width standards; standard artefacts for calibration of e-beam and X-ray production equipment etc.

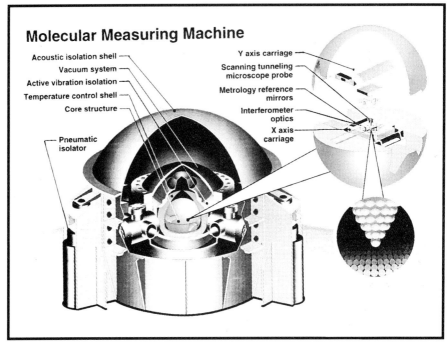

Figure 12   Schematic of the NIST Molecular Measuring Machine
(Courtesy of National Institute of Standards and Technology, USA)

The maximum size of the specimen to be measured is 50 mm x 50 mm x 25 high. It employs very high resolution scanning tunnelling and atomic force microscopes (STM and AFM). The target resolution of point-to-point measurement within the measuring volume was 0.1 nm (1Å). However, the X and Y interferometers that measure displacement with respect to the X and Y ultra precision Zerodur metrology reference mirrors have been improved to operate at a resolution of 0.05 nm i.e. 0.5Å or about one-fifth the distance of typical inter atomic spacings. The target of 0.1 nm accuracy over 50 mm explains the extraordinary care that has had to be taken in the design to isolate it from environmental effects such as seismic vibration (the structure having a first resonant frequency of 2 Khz and active vibration isolation), acoustic vibration (acoustic isolation shell) temperature changes (temperature control shell), humidity, CO_2 content, temperature (vacuum system, 10^{-7} Pa).   Control of temperature is maintained to within 0.0005°C (0.5 mK).   By measuring single crystal surfaces with known atomic order and spacings, the machine can be calibrated against an atomic length standard; software error compensation can be used to virtually eliminate the residual dimensional measurement errors of this state-of-the-art machine.

C527/

6. SOME NATIONAL AND INTERNATIONAL INITIATIVES IN MICRO AND NANOTECHNOLOGY

6.1 The UK

The DTI launched a LINK Nanotechnology Programme in 1988 which was joined by SERC in 1989. About £12.5 m was put in by government agencies and this was supplemented by over £16 m of industrial funding. The programme supported 29 projects covering ultra-precision machining, nanopositioning and control, nanometrology, surface characterisation and analysis, ultra fine powders/particles, biomedicine/biosciences and micro mechanics. It is considered to have been a major success in industry/science base, multi-disciplinary research. Since 1994, the EPSRC has also supported a managed programme in nanotechnology with finance of approx. £3 m; all projects are due to be completed by 1999 (8).

6.2 Europe/EU

As yet there are no formal nanotechnology programmes funded by the European Commission (EC) although there are several programme areas involving micro and nanotechnology. Protein engineering and biosensors, etc. are included in the Biotechnology Action Programme. The current BRITE/EURAM programme in Materials Science have identified 17 projects which involve nanotechnology as well as others in the area of 'microtechnology'. Nanotechnology is also a part of the Biomedical Technology Programme, e.g. artificial limbs, nerves, hearing, sight etc.

Within the current 4th Framework Programme, running from 1994 to 1998, support is generated to **Microsystems Technology** (9). The EC has also made available about £20 m for an ESPRIT Advanced Research Initiative in Microelectronics.

Networks funded by the EC also include:
(i) PHANTOMS which is co-ordinating work on Mesoscopic Physics and Technology. The research covers the areas of nano-electronic components, nanotechnology fabrication, opto-electronics and novel architectures for switching. There are 18 countries in this programme.
(ii) NEOME (Network Excellence on Organic Materials for Electronics) has run since 1992 and is co-ordinating work in the area of organic materials for electronics as the basis for nanostructures, optical devices, electro-optical switches, sensors and for potential applications in information storage and information processes. Ten countries are currently directly involved in this network.
(iii) NEXUS (European Network of Excellence in Multi-functional Microsystems) was established in 1992. Its brief includes: a) co-ordinating work in microsystems technology (MST), b) the strategic assessment of international MST developments, and c) the formation of links to small and medium sized enterprises.
(iv) EUROPRACTICE which has been established to stimulate the exploitation of state-of-the-art microelectronics technologies by European industry to improve industrial competitiveness.
(v) NANO to co-ordinate research and development in nanostructural materials.

A proposal is being submitted to the EC during 1997 for the establishment of the European Society for Precision Engineering and Nanotechnology (EuSPEN). The Society aims to strengthen industrial and research potential in precision engineering and nanotechnology by establishing a society, organising events, initiating transnational research projects, building a database and promoting education and training. The EuSPEN

Foundation Committee comprises at least one academic and one industrialist from 6 countries, namely Denmark, France, Germany, Italy, The Netherlands and the UK; it will cover most EU countries within four years.

6.3 The USA

Progress in nanotechnology is being pushed in the main by industry, and in particular the larger firms. However Government funding bodies supporting Micro-Electro-Mechanical Systems (MEMs) are a) the Advanced Research Projects Agency (US$ 40 million per annum), b) the National Science Foundation (US$ 5 million per annum), and c) the National Institute of Standards and Technology (NIST) (US$ 5 million per annum). In addition the US Department of Defence Initiative on MEMs funded special 'sensitive' projects with US$ 46 million in 1996 and this has risen to US$ 63 million in 1997.

The main activities in USA micro and nanotechnology programmes are:
(i)   MEMs which encompasses all microsystems technologies and devices
(ii)   Nanofabrication for manufacturing devices with nanometre scale structures, and
(iii) Aspects of the traditional science base (e.g. physical, chemical and biology) appropriate to 'nanoscience', including biotechnology.

A National Nanofabrication User Network (NNUN) has been established. Successful projects in the areas of biology, chemistry, medicine and physics are presented in appropriate forums. The NNUN has a hub of 2 user facilities available, one on the east coast (Cornell University) and one on the west coast (Stanford University). In addition three additional sites offer expertise in the specific areas of a) wide band semiconductors (Howard University), b) novel materials (Pennsylvania State University, and c) semiconductor etching (University of California Santa Barbara). NIST is a leading activity in nanometrology and nanotechnology.

6.4 Japan

The importance of nanotechnology in Japan can be judged by the large amount of government funding it attracts. Funding is made available by 3 separate ministries: Ministry of Trade and Industry (MITI), the Science and Technology Agency (STA), and the Ministry of Education.

In 1992 MITI started a major 10 year project **'Research and Development of Ultimate Manipulation of Atoms and Molecules'** with total government funding of the order of US$ 250 million. The research consortium currently includes 26 Japanese and 4 foreign companies and the broad range of applications expected to emerge from the programme range from computation (e.g. higher density computer memory), to new materials, gene manipulation and new catalysts for environmental clean up.

A further large 10 year project started in 1991, funded by MITI to over US$ 200 m, is the NEDO **Micromachine Technology**. Twenty eight companies and institutes are involved in the programme which is undertaking research in sensors, actuators, energy supply and system control. It plans to encompass "micro-factories", fully functional micro manufacturing systems comprising very small machines grouped within table top dimensions for the manufacturing of very small industrial products such as micro pumps, micro valves, micro motors and other micro machines.

STA projects have focused on various aspects of nanotechnology over recent years, including **Nano-Mechanisms** (1985 to 90), **Solid Surfaces** (1985 to 90), **Molecular Dynamic Assembly** (1986 to 91), and the **Atom Craft** project (1989 to 94) which is investigating the behaviour of atoms and molecules on surfaces together with techniques for precision deposition, based upon the application of STM technology. STA is also responsible for the Institute of Physical Chemical Research (RIKEN), which ran a 'Frontier Research Programme' in the areas of molecular electronics, bioelectronics, and quantum electronics. One objective of this programme is the development of an 'artificial brain'.

7.   CONCLUSIONS

Manufacturing technology is at the start of dramatic changes as micro and nanotechnology products and processes expand and, in some cases, usurp conventional macro manufacturing. The commercial opportunities of the so-called "Nanometre Age" (10) are broadly shown in Fig. 13.

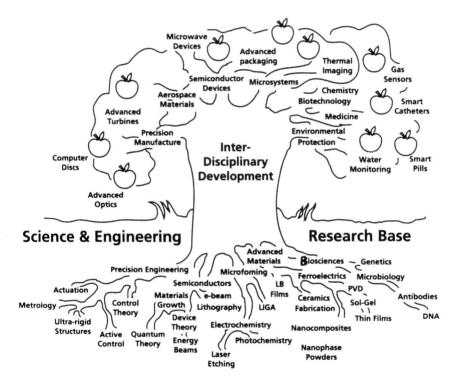

Fig. 13   The Nanotechnology Tree of Commercial Opportunities
(Courtesy Professor Roger Whatmore, Royal Academy of Engineering
Professor of Nanotechnology, Cranfield University)

The world market size for micro and nanotechnology products is likely to be very large; a few reports give estimates in excess of US$ 400 bn by 2002 AD:

- a 1995 VDI report for the German Government estimates that by the year 2000 it will exceed US$ 250 bn (11).
- the market for MST/MEMS alone will exceed US$ 14 bn by 2000;

- the market for 0.1 µm (100 nm) DRAMS will exceed US$ 100 bn also by 2000.

Micro and nanotechnology is a major new technological force that will have substantial socio-economic effects throughout the world. Many benefits in standards of living and quality of life can be confidentially expected.

8.   REFERENCES

1.   McKeown, P.A. "High Precision Manufacturing and the British Economy", James Clayton Lecture, Inst. of Mech. Engrs., Proc. 1986, Vol 200, No 76.
2.   Taniguchi, N. "On the basic concept of Nanotechnology", Proc. Int. Conf. Prod. Eng., Tokyo, JSPE 1974.
3.   Taniguchi, N. "Current status in and future trends of ultra precision machining and ultra-fine materials processing". Annals of the CIRP, 1983, 32(2).
4.   "Nanotechnology, integrated processing systems for ultra-precision and ultra-fine product," edited by N Taniguchi, ISBN 0 19 8562837 published by Oxford University Press, 1996.
5.   Feynman, R. "There's Plenty of Room at the Bottom" : An Invitation to Enter a New Field of Physics", Engineering and Science Magazine Feb. 1960; California Inst. of Technology, USA.
6.   Drexler, K.E. "Engines of Creation : the Coming Era of Nanotechnology", Doubleday, 1992.
7.   McKeown, P.A. Course notes on Precision Engineering, School of Industrial and Manufacturing Science, Cranfield University.
8.   Parliamentary Office of Science and Technology, London SW1P 3JA; report. "Making it in Miniature - Nanotechnology, its applications and UK Science" - issued to Parliament in October 1996.
9.   Dorey, H., Yeatman, E.M. 1995 "Micro Systems Technologies in Europe Today - a study on the actual situation in European Companies". ESPRIT 8520, EU. Edited Rave. (Professor Howard Dorey, F.Eng. (Microengineering), (Imperial College of Science, Technology and Medicine, London SW7 2BT).
10. Rohrer, H. "The Nanometre Age : Challenge and Chance", Micro Electronic Engineering 27 (1995) 3-15, Elsevier Science BV.
11. VDI - Technologiezentrum Physikaliche Technologien, 1995   "Analysis and Valuation of Future Technologies: Technology Analysis - Nanotechnology", for the Federal Ministry of Research and Technology, Dusseldorf, 1995, p. 108.

crofabrication – from invention to innovation

ssor Dr W EHRFELD
ging Director, Institute of Microtechnology Mainz GmbH, Germany

ABSTRACT

Microfabrication is based on precision engineering and semiconductor fabrication technologies, extending the successful concept of miniaturization from microelectronic circuits to mechanical, optical, fluidic, chemical and other functional elements. A variety of microfabrication technologies has been developed in the last two decades which allow the production of threedimensional ultra precise components without major restrictions in design and materials. Bulk- and surface micromachining of silicon, laser-based processes, the LIGA technique, micro spark erosion and micromilling as well as a number of etching processes have led to a broad spectrum of new products covering nearly all sectors of modern technologies. The paper deals in particular with the LIGA technology, which combines an extremely high precision with a wide variety of materials and the capability for a cost effective mass replication. This technique has marked a way from invention to innovation in microtechnology by creating a broad basis for advanced products and new markets.

1 INTRODUCTION

The unparalleled success of microelectronics has shown that miniaturization offers a variety of advantages ranging from saving of material, space and energy over cost saving by use of cost efficient batch fabrication technologies to the dramatic increase of performance by integration of functional elements. Microtechnology or microsystem technology is commonly regarded as the consequent application of miniaturization as a general strategy of success, not only for electronic components and systems but also for a huge variety of mechanical, optical, acoustic, thermal, fluidic, chemical or biochemical functional units.

The promising perspectives of microtechnology have led to a considerable raise of public funds and to a broad public attention often generating confusion by mixing up visions with the state of the art in microfabrication. It is a common impression that microsystems will be successful in a few years but did not find any commercial application field at present. However, comparing science fiction to reality one finds that microsystems nowadays have a multi billion $ presence on the market and huge yearly growth rates: today products are found in the field of communication technology like read/write heads of hard disks or ink jet printer heads, in automotive applications like airbag sensors and navigation systems, in a variety of consumer products like recording heads for CD-players or microchips with some hundreds of thousands mirrors for high resolution projection tv, and in medical applications like in endoscope or catheter systems for minimally invasive diagnostics and therapy.

Recently fabricated microproducts bear a high resemblance to subassemblies and elements of traditional precision engineering. In fact, microsystem technology is based both on precision engineering and semiconductor fabrication technologies, using various fabrication technologies for the production of a multitude of functional elements. A broad spectrum of materials is required for the realization of this functional variety and the use of different fabrication technologies for all these components is the approach of microengineering, which is the basis for the development of innovative systems for nearly all fields of modern technology. By now, the development of fabrication technologies has reached a state where industrial production of threedimensional microcomponents is feasible almost without restrictions in design and materials.

Techniques which mainly evolved from microelectronics, e.g. wet-chemical anisotropic etching of single-crystalline silicon (bulk micromachining), or dry-etching processes by means of low-pressure plasma or ion beams (surface micromachining), were among the first used in new disciplines like micromechanics and microoptics. Today, the commercial success of these technologies is beyond any discussion, in particular for the huge application field of miniaturized sensors. They suffer, however, from restrictions in the choice of proper materials and comparatively high material costs and, as a result, they are not applicable when metals, ceramics or plastics are required. Other microfabrication technologies e.g. structurization of photosensitive glass, laser micromachining or mechanical micromilling or micro spark erosion have also specific advantages and drawbacks regarding e.g. materials, precision or production costs.

Among all microfabrication technologies existing to date, the LIGA process offers a particularly broad spectrum of design possibilities when regarding precision, threedimensionality, materials and cost saving by mass fabrication through replication processes. Development work on LIGA has started worldwide and the period of pure invention has changed into the period of innovation. The scope of this contribution is the presentation of the state of the art of the fabrication process as well as the description of some promising fields of application for LIGA technology and precision engineering.

2   THE LIGA MICROFABRICATION TECHNOLOGY

The LIGA technology [1,2] is based on a sequence of process steps combining deep lithography, microelectroplating and micromoulding (German acronym: Lithographie, Galvanoformung, Abformung) (see Fig. 1). In the first step, a threedimensional structure is generated by means of a lithographic process in a thick resist layer. This can be achieved by utilizing a laser, an electron or ion beam, as well as by standard optical- or by means of X-ray lithography which has been shown to give particularly good results. In the second step, a complementary metal structure is generated from the resist master by means of electroforming. This metal structure serves as a mould insert for subsequent replication processes like injection moulding, reaction injection moulding, embossing or slip casting. The replication of the metal master allows low cost mass fabrication of threedimensional microstructures using a wide variety of materials ranging from metals, metal alloys and ceramics to polymers. In addition, metallic

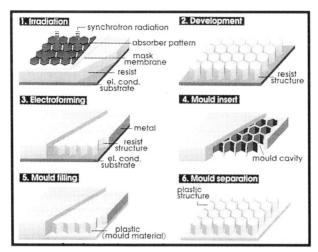

Fig. 1: Typical sequence for the production of microstructures using the LIGA-process.

LIGA microstructures can be used as microelectrodes for spark erosion which increases further the spectrum of applicable materials like stainless steel, titanium, shape memory alloys and electrically conductive ceramics.

The simple shadow printing process of deep X-ray lithography results in relief structures with straight parallel walls. Complex threedimensional structures have obtained by means of an advanced modification which is based on aligned multi exposure techniques in combination inclined irradiations. IMM and JENOPTIK GmbH have developed a special X-ray scanner system, that is particularly suited for these purposes [3]. It offers high precision mask alignment for multi exposures and inclination as well as rotations of the mask and substrate with respect to the X-ray beam axis. Fig. 2 gives examples of microstructures that have been realized with this scanner system by double exposure (left) and by a combination of multi exposure and inclined irradiation at different angles (right) [4].

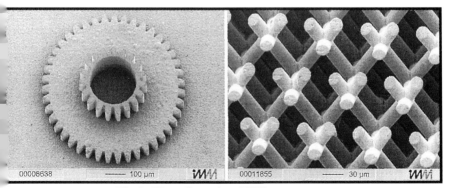

Fig. 2: Threedimensional LIGA structures.
Left: stepped gear wheel generated by aligned double exposure and electroforming from nickel.

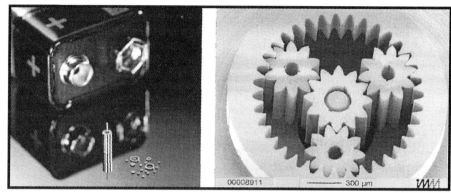

**Fig. 3: Left: Assembled permanent magnet micromotor with integrated gear system
Right: SEM-view of the micro gear box, material: NiFe.**

3 FIELDS OF APPLICATION AND EXAMPLES OF USE

The development of LIGA process technology has been accompanied from the very beginning by intensive R&D work on LIGA microdevices. Thus, a way has been opened which leads from invention to innovation, i.e. to the commercial utilization of the LIGA process for mass fabrication of microcomponents and microsystems. Meanwhile, LIGA products are under development for applications in information and communication technology, chemical engineering, molecular biotechnology, medical technology, process automation and many more. Dispensing systems, microspectrometers, ink jet printer heads, optical fibre ribbon connectors, micro reaction devices, micro filtration systems, micropumps and micro gear systems have been introduced or will be shortly introduced to the market. In addition, special equipment for the LIGA process like X-ray scanners and electroforming apparatus has been developed and commercialized worldwide. In the following, some examples of use will be described in more detail.

3.1 Permanent magnet micromotor with integrated gear system

Micromotors are particularly interesting for applications medical technology, e.g. for novel catheter and endoscope systems in minimally invasive diagnosis and therapy, as well as in micropositioning applications and drive systems for consumer products. In most cases, such micromotors have to be equipped with a gear system in order to obtain a sufficiently high torque.

Fig. 3 shows such a motor with an integrated gear system [5,6]. While the motor can be produced by standard precision engineering, the gear system must be fabricated by LIGA technology in order to obtain sufficiently low tolerances. The motor has been developed in close cooperation with the company Dr. Fritz Faulhaber GmbH. With a diameter of 1.9 mm and a length of 4 mm, a motor torque of 5 µNm is achieved at a rather low voltage of 1.5 V which makes this motor particularly interesting for medical applications. The maximum revolution speed exceeds 100.000 rpm. No failures have been detected in life time tests of several months with 10.000 rpm. Serial production of the permanent magnet motor is presently starting.

The design concept of the IMM-micromotor (see Fig. 3) is based on the use of permanent magnets which constitute a powerful source of magnetic energy. The synchronous motor

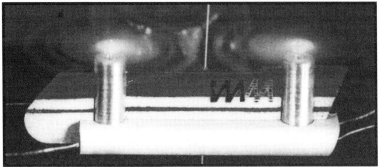

Fig. 4: The smallest helicopter of the world

scheme leads to a low power consumption and small heat production. The rotor consists of a tiny steel shaft and a diametrically magnetised rare earth magnet (Fig. 3). Two bearings, fixing the rotor shaft on both sides, are fastened to a housing tube made from soft magnetic material, that generates a closed loop for the magnetic flux.

The high rotational speed of the motor can be converted by use of micro gear boxes to increase the torque for specific applications. Therefore, a separate gear box has been developed which fits to the motor dimensions. The design is based on an epicyclic gear system (see Fig. 3) in order to meet the rather strong demands on compactness and high torque which are met due to a transmission rate of 45. The modified Wolfrom type gear unit consists of 300 µm high planet and sun gears and of two internal toothed hollow gears with a height of 600 µm. One of the hollow gears is fixed with respect to the motor housing, while the other is used as output drive. LIGA technology is applied to fabricate gear wheels from either metal or polymer materials. The replicated structures exhibit an extremely high precision which eases the assembly process. Further on, the extremely low mean wall roughness of R_a = 40 nm leads to very low friction losses and a smooth and precise gear roll off. The reliability of the micromotor with the integrated micro gear box has been proven by continuous operation in a number of laboratory set-ups.

To demonstrate the huge range of applications and the high power of the permanent magnet micromotor, IMM built the world's smallest helicopter shown in Fig. 4 [6]. It consists of two micromotors fixed to an aluminum body. Gear units are not applied in the present case. Thin wires are used to supply the engines with power. With a length of 24 mm, a height of 8 mm and a weight of 400 mg the helicopter takes off at 40.000 rpm of the contrarotating rotors.

The helicopter is an object which shows the efficiency of the micromotors. But it also demonstrates that microtechnology is not only science or pure engineering. The application shown above therefore presents a new and fascinating opportunity for microtechnical products: the fabrication of toys.

3.2 Optical microconnectors

The huge growth rates in the market of optical fibre communication urges the development of ultra precise positioning and interconnection devices. It has been shown in various applications that LIGA technology offers a spectrum of elegant solutions for optical communication technology. Part of the work performed in this application field is summarized below.

A fibre ribbon connector has been developed that is capable of aligning ribbons of 12 singlemode fibres with an absolute precision of better than 1 µm. The design allows to balance out tolerances in the diameter of the fibres, thus minimizing coupling losses. The connector is based on elastical polymer V-grooves that grip the ends of the fibre in a "teeth-like" way (see

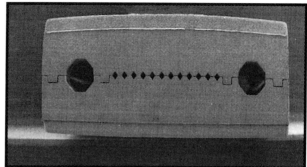

Fig. 5: Front face of the fibre ribbon connector showing alignment structures and guide pin holes. Fibre pitch is 250 μm.

Fig. 5). The mould insert for this ferrule is a combination of precision nickel parts from the LIGA process and of steel parts with less precision fabricated via electro discharge machining.

For fibre-in board technology to be used in ultra fast data transfer rates, metallic and platic comb structures have been realized that are capable to align ribbons with 16 multimode fibres emerging from the board under an angle of only a few degrees. This alignment structure is combined with a second snap in LIGA part which allows to control precisely the finishing process of the fibres (see Fig. 6).

The counterpart for the fibre alignment structure, e.g. an emitter or photodetector array, is pigtailed and the fibre ends are positioned in the same way using a corresponding comb element. The alignment of the device relatively to the board fibres is achieved using guide pins mounted in the cable-side device.

3.3 Fibre optical switch

An example for a microactuator suitable to switch the light of an incoming fibre to two alternative outgoing fibres is given in Fig. 7. Incoming and outgoing fibres are mounted on a soft magnetic NiFe-slide which is guided by integrated parallel springs. Switching is performed with short current pulses in the coil pairs, thus closing the magnetic path of the NiFe-structures and generating sufficiently high forces to move the slide to the desired direction [8].

For the fabrication of these fibre switches, LIGA technology has been combined with the sacrificial layer technique. This type of switches has been tested extensively and revealed switching times of less than 1 msec and an extremely high precision at the two end positions

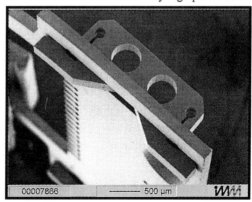

Fig. 6: Alignment part for fibre-in-board technology.

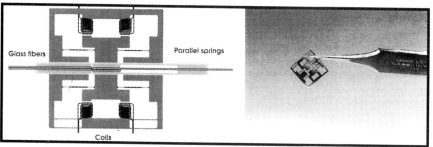

**Fig. 7: Left: Schematic view of an optical fiber switch.
Right: SEM view of the fiber optical switch before assembly.**

In the schematic: Glass fibers, Parallel springs, Coils

of the carriage. No sign of fatigue is observed for the springs after more than 10[8] oscillations. At present, further generations of such fibre-optical switches are under development which will allow to realize NxM matrix switches with more than 100 fibres for the rows and the columns of the matrix.

3.4 Components for microreaction technology

Microreaction technology [9,10,11,12] is currently regarded as one of the most promising perspectives for chemical engineering, combinatorial chemistry and molecular biotechnology. Micromixers, micro heat exchangers, units for phase transfer, reaction chambers, intelligent fluidic control elements and microanalytical devices allow to perform all unit operations of chemical engineering in an integrated miniature system. Because of the small dimensions, such systems have an extremely high surface to volume ratio and very short response times. This results in a variety of advantages, e.g. the feasibility of novel process routes, the reduction of reaction byproducts, the increase of yield and selectivity and a high production flexibility. In addition, new devices can be realized to synthesize and analyze a huge number of substances for mass screening of chemical and biological substances.

The progress in microreaction technology is directly related to the availability of suitable microfabrication technologies for chemically inert or biocompatible materials, e.g. metals, metal alloys, polymers, ceramic materials and glasses. Hybrid concepts are required as well as the combination of LIGA with conventional precision engineering technologies.

Some typical examples of microreactor components are shown in Fig. 9. The plate-type micro heat exchanger given in Fig. 8 (top, middle) is an illustrative example for the performance of a microreactor component. It allows to transfer some 10 kW/cm^3 of active volume under countercurrent flow conditions. Such a heat exchanger plate fabricated by means of the LIGA process has been transferred into hard metal by micro spark erosion. This hard metal structure has been used as a tool to manufacture large numbers of heat exchanger plates from aluminium by means of hot embossing. A modification of this process allows to fabricate heat exchanger plates from ceramic materials for applications in highly corrosive atmospheres and at high temperatures, respectively.

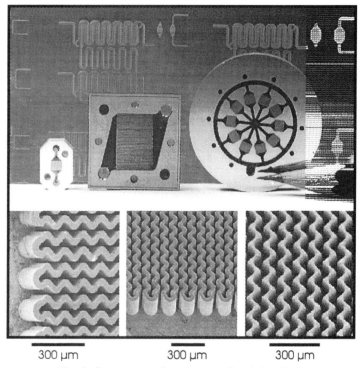

Fig. 8: Components for microreaction technology.
Top: Micro mixers (left, right) and micro heat exchangers (middle).
Bottom: Flow channels of static micro mixers made from nickel (left), silver (middle)
and titaniumdiboride (right).

Micro mixers as shown in Fig. 8 usually bring together two streams of fluids in countercurrent flow. These streams are subsequently separated in thin alternating layers, which are mixed within a short distance. This procedure can be repeated or enhanced by using corrugated ducts (see lower part of Fig. 9). The microstructures shown there are fabricated from nickel, silver and titaniumboride, respectively, which demonstrates the wide variety of materials applicable in LIGA based microfabrication. By means of such static mixers also a very efficient enlargement of the surface area can be realized for applications in phase transfer catalysis, which results in extremely high conversion rates per unit volume.

Micropumps as shown in Fig. 10 combine the ability of dosing very small amounts of liquids with low manufacturing costs and small sizes [13]. They can be used in many applications ranging from medical technology (e.g. drug delivery, infusion, disposable dosing systems) and chemical engineering (e.g. process control, micro reaction technology) to more specialised tasks (e.g. hand held equipment, miniaturized analysis systems).

The components of the membrane micropump (left side of Fig. 9) have been fabricated by means of micro injection moulding using mould inserts generated by deep X-ray lithography and electroforming. Assembly and connection is done by laser welding using integrated positioning structures for alignment. The self-priming membrane micropumps which are equipped with piezo actuators deliver flow rates up to 300 µl/min, while the maximum back pressure exceeds 1000 hPa. The pump design is suitable for mass production.

Fig. 9: Left: Membrane micropump on top of a microtiterplate.
Right: Micro gear pump.

Micro gear pumps (right side of Fig. 9) are particularly useful for highly viscous fluids. These pumps are also self-priming and may achieve back pressures of some bars. Gear wheels of standard LIGA gear systems can be applied which allow to realize low manufacturing costs.

4 CONCLUSION

The development of the LIGA process and the corresponding fields of application demonstrate that modern microfabrication technologies can be regarded as a consequent continuation of classical precision engineering. Meanwhile, microtechnology has reached a status of nearly unlimited design flexibility combined with an extremely broad spectrum of applicable materials.

Most examples described in this contribution show that micro systems are no monolithic devices like microelectronic circuits. The realization of microsystems is rather based on assembly and interconnection of micro components where the assembly tools are manufactured to a large extent by means of microfabrication methods. In this respect, microtechnology corresponds largely to precision engineering.

In many fields of modern technologies, micro structure products are starting to take over the role of standard products of precision engineering. The period of invention in microtechnology has changed into the period of innovation and commercialization.

5 REFERENCES

[1] W. Ehrfeld, H. Lehr; Radiat. Phys. Chem. 45, No. 3 (1995), pp. 349 ff

[2] W. Ehrfeld, H. Lehr; Synchrotron Radiation News 7 (5), pp. 9-13, 1994

[3] A. Schmidt, A. Clifton, W. Ehrfeld, G. Feiertag, H. Lehr, M. Schmidt, Microelectronic Engineering, 30, pp 215-218, 1996

[4] G. Feiertag, W. Ehrfeld, H. Freimuth, H. Lehr, M. Schmidt, R. Weiel; Proc. of Int. Symp. on Microsystems, Intelligent Materials and Robots, Sendai, pp. 37-40, 1995

[5] K.-P. Kämper, W. Ehrfeld, B. Hagemann, H. Lehr, F. Michel, A. Schirling, Ch. Thürigen, T. Wittig, Proc. Actuator 96, Bremen, 1996

[6] H. Lehr, S. Abel, J. Döpper, W. Ehrfeld, B. Hagemann, K.-P. Kämper, F. Michel, Ch. Schulz, Ch. Thürigen, Proc. SPIE's Photonics East Symp. Boston, 1996

[7] H.-D. Bauer, W. Ehrfeld, M. Gerner, T. Paatzsch, A. Picard, H. Shift, L. Weber, Proc. Int. Symp. Microsystems, Intelligent Materials and Robots, Sendai, pp. 33-36, 1995

[8] S. Abel, W. Ehrfeld, H. Lehr, H. Möbius, F. Schmitz; Proc. MicroMat '95, Berlin (1995)

[9] W. Ehrfeld, V. Hessel, H. Möbius, Th. Richter, K. Russow in: Microsystem Technology for Chemical and Biological Microreactors (W. Ehrfeld, Ed.), DECHEMA Monograph, 132, VCH, Weinheim, 1996

[10] W. Ehrfeld, K. Golbig, V. Hessel, H. Löwe, Th. Richter, K. Russow, mst news, 17, p. 5, 1996

[11] W. Ehrfeld, GIT special edition on chromatography, 2, p. 83 ff, Darmstadt, 1996

[12] W. Ehrfeld, K. Golbig, V. Hessel, R. Konbrad, H. Löwe, Th. Richter, Proc. 1. Int. Conf. on Microreaction Technology, Frankfurt, 1997 (to be published)

[13] J. Döpper, M. Clemens, W. Ehrfeld, K.-P. Kämper, H. Lehr; Proc. ACTUATOR 96 (1996); pp.37-40

crosystems engineering

essor **R A LAWES** BSc, FCGI, FInstP, FIEE, FEng, FRAEng
, Central Microstructure Facility, Rutherford Appleton Laboratory, Didcot, UK and Visiting Professor of
rical Engineering, Imperial College, London, UK

1.   INTRODUCTION

It is timely that, on the 150th Anniversary of the Institution of Mechanical Engineers, mechanical engineering should be entering a period of major innovation.   The traditional methods of processing, forming and assembling mechanical components that have been greatly refined over the last 150 years are now being superseded in many areas by a new engineering process often referred to as *Microsystems Technology* or microelectromechanical systems.

Microsystems Technology (MST) is based upon techniques established by the semiconductor industry to manufacture the modern integrated circuit.   The integrated circuit, particularly in the form of the "silicon chip" has formed the basis of entire modern industries such as computers, mass communication, video games. Application specific integrated microsystems may well create a similar industry by the early part of the 21st century.

Microsystems technology can combine electronics, mechanics and optics at significantly reduced dimensions to achieve high performance and high complexity in a very small volume, eg a few cubic millimetres.   Thus, MST enables unique products to be made that would be either too expensive or impossible to manufacture with conventional "macro-engineering" techniques. For example, drug delivery systems are being developed using MST, that will be small enough to be implanted into the human body, containing sensors, control valves, an intelligent computer chip and a power source.   Such systems will deliver minute controlled amounts of a particular drug and function autonomously for many years.

Most areas of engineering are starting to benefit from MST, either through the development of novel products or through improvements to existing products, eg the airbag crash sensor. Consumer electronics, mobile communications, automotive/traffic control, automation, aerospace, medical, environmental and the household have been identified as markets that will be revolutionised by MST products.

2. THE NEW MICROFABRICATION TECHNOLOGIES

MST still requires similar supporting technology to macroengineering, for example, design, modelling and simulation, new materials, new assembly methods and testing. However, the key to MST is the microfabrication technology, initially adapted the from semiconductor industry but now being developed in new directions.

Much of the pioneering fabrication technology for MST has been based upon silicon and hence silicon processing manufacturing techniques. Silicon process compatible materials such as SiO_2, Si_3N_4 and poly-silicon have been widely used, along with conventional silicon processes such as photolithography, ion implantation, chemical vapour deposition (CVD) and reactive ion etching. However, the development of MST will require a wider range of materials to be available, eg ceramics, plastics and consequently new microfabrication techniques will need to be employed. Structures 2-3 orders of magnitude thicker than have been achieved with current silicon technology will be required if many electromechanical concepts are to be realised.

2.1 Bulk Silicon Micromachining

One of the first attempts to use some of the fabrication techniques commonly used in the microelectronics industry has been with *bulk micromachining*.

The technique relies upon the selective etching (usually wet chemical) of single crystal silicon either by using an etch stop layer or by exploiting the preference of chemicals to etch certain silicon planes and not others. This relatively simple technology enables microcomponents such as membranes, microcantilevers and microactuators to be built.

For example, ethylenediamine pyrocatectol (EDP) will etch silicon but not SiO_2, Si_3N_4 or heavily boron doped silicon. Hence by boron doping the top surface of a silicon wafer, protecting the outer region of the back of the wafer with a SiO_2 masking layer or a metallic ring and immersing the wafer in EDP, membranes a few microns thick and several centimetres square can be manufactured. Fig 1(a) shows a number of membranes on a wafer fabricated in such a manner.

Fig 1(b) shows a <100> silicon wafer which has been coated with a thin layer of SiO_2. The masking pattern shown is etched into the SiO_2 and aligned to the <110> direction on the surface of the wafer. Using EDP as the anisotropic, selective etch, the <111> planes will etch more slower than the others resulting in the flat-bottomed

V-groove structure shown. If etching is continued then the silicon under the central SiO₂ structure will be removed leaving an SiO₂ cantilever in the <110> direction suspended over a pit with sides in the <111> direction as shown in Fig 1(c). Such microcantilevers with a seismic mass attached and piezoelectric stress sensors integrated by lithographic techniques, form the basis of a high performance, rugged and cheap accelerometer.

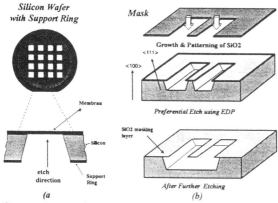

FIG 1 Bulk silicon microengineering using (a) boron doped silicon as an etch stop and (b) preferential etching of Si crystal planes using SiO₂

Fig 2 shows a microfluidic inlet manifold for a blood analysis instrument, where the manifold has been constructed using SiO₂ as the mask. Fig 3 shows part of a comb microactuator manufactured using the boron-doped silicon etch stop technique.

FIG 2 A microfluidic manifold
(Courtesy Univ of Hertfordshire)

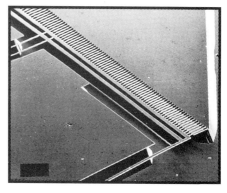

FIG 3 A microactuator manufactured
from boron-doped silicon (Courtesy of
Imperial College, London)

2.2   Surface Micromachining

Bulk silicon micromachining has considerable limitations which constrain the designer to a relatively simple set of mechanical possibilities. A direct extension of semiconductor manufacturing technology known as *Surface micromachining* enables structures to be manufactured that are at least an order of magnitude smaller than those from bulk micromachining. As the wafer surface containing the mechanical structures is also available for the microelectronic components, surface micromachining enables microelectronic and micromechanical components to be integrated on a single silicon chip.

The key to surface micromachining is the sacrificial layer, analogous to the "lost wax" process used in metal casting technology. There are a variety of materials that can be used to fabricate the required artefact and to provide the sacrificial layer. CVD silicon can be used to form the mechanical structure with SiO_2 as the sacrificial layer. SiO_2 may be deposited by CVD or grown thermally. Fig 4 shows the essential steps to produce the micromechanical structure, namely to deposit the sacrificial layer, to deposit the microstructures and to remove the sacrificial layer by an isotropic etch. Another technique uses photoresist as the sacrificial layer, which can be readily removed with acetone or a photoresist developer, and nickel as the structural material. A comb microactuator using this method and manufactured on a glass substrate is shown in Fig 5.

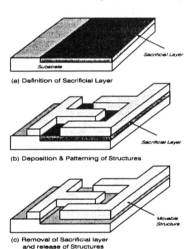

(a) Definition of Sacrificial Layer

(b) Deposition & Patterning of Structures

(c) Removal of Sacrificial layer and release of Structures

FIG 4 Surface micromachining using a sacrificial layer

FIG 5 Free standing nickel microactuator fabricated using a photoresist sacrificial layer

The basic surface microengineering techniques described above have been used to manufacture many complex micromechanical systems. For example linear motion has been achieved using a comb-actuator similar to that shown in Fig 5. The comb-actuator consists of a number of interdigitated electrodes that electrostatically generate force and a suspension providing guidance to the resultant linear movement. The actuation force is proportional to the number of pairs of electrodes, the thickness of the electrodes, the dielectric constant of the air gap and the square of the voltage applied. It is also inversely proportional to the gap between the electrodes and hence the gap is only a few microns. Typically a force of 10 nanonewtons is produced. Pairs of actuators can be combined to produce a X-Y microstage capable of positioning to 50 nm. Such microstages are an important contribution to the field of precision 2D-positioning and offer a cheap, mass producable alternative to servo motor or piezoelectric driven microstages.

Electrostatic motors have been fabricated with rotors in the order of 100 μm diameter and a few microns thickness. For a typical 12:8 motor (6 stators and 4 rotor poles), as shown in Fig 6 a force of 0.27 micronewtons can be generated at the pole of a motor where the thickness of the motor t = 2.5 μm, the stator-rotor gap = 2 μm and the stator-rotor voltage V = 200 Volts. Thus a motor with a rotor diameter of 100 μm can generate a torque of 22 piconewton-metres and generate approximately 20 nanowatts of power at 10,000 rpm.

FIG 6 A 12:8 electrostatic motor manufactured by surface microengineering.
Voltage is supplied sequentially to the pole pair as shown at a frequency equal to
the rotational speed required from the motor

Currently motor lifetimes are short and bearing friction forces high. Considerable improvements to bearing design and the use of new materials and the fabrication of much deeper microstructures are essential if micromotors with usable torque, power and lifetimes are to be achieved.

2.3   Deep Microstructures

The forces available increase with the thickness of the device and so technologies that are capable of making structures 100-1000 μm high, while maintaining micron or submicron resolution in plan section, are highly desirable. One such technology is known as LIGA (Lithographie Galvanoformung, Abformung) based upon deep etch lithography, electroplating and optional moulding processes.

LIGA is similar to X-ray lithography which has been developed for the semiconductor industry to print features close to and below the diffraction limit of light optics, eg. ≤0.25 µm. In such a system a mask, consisting of a patterned gold absorber on a thin substrate, is illuminated with soft x-rays of about 1 nm wavelength and projected on to a resist-coated wafer. LIGA, as used for micromechanics, requires much harder X-rays of about 0.1 nm (12 keV) wavelength to penetrate the much thicker resist (eg 100 - 1000 µm deep). Electron accelerators known as synchrotrons produce "synchrotron radiation" as a result of the deceleration of electrons in a circular orbit and can produce intense beams of hard X-rays able to penetrate the full depth of the resist with very little lateral exposure due to the photoelectrons released in the resist. Relatively thick membranes (100 - 200 µm) of beryllium or diamond can be used as the supporting mask substrate and a 20 µm layer of gold as the mask absorber.

The basic steps of LIGA are shown in Fig 7 starting with the exposure of a deep resist coated substrate to hard X-rays (Fig 7a). Subsequent resist development will produce the structure shown in Fig 7b and electroplating will produce a reversed metal structure as shown in Fig 7c. After resist removal this metal version can be used as the microstructure itself or act as the mould for a specially developed injection-moulding process to make thermoplastic or resin microstructures (Fig 7d). Alternatively, by injecting the mould with a suitable slurry, ceramic microstructures can be fabricated.

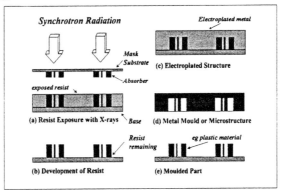

FIG 7 The basic steps of LIGA showing X-ray exposure, electroplating and moulding

The LIGA technology offers considerable advantages when compared with silicon microengineering. Firstly, it meets the need for deep structures particularly where high aspect ratio microstructures are needed with vertical sides. For example electrostatic micromotors need to be as deep as possible with micron-sized stator motor gaps to maximise torque. Secondly it greatly extends the range of materials that can be used in microengineering other than silicon and related materials eg metals, plastics, ceramics.

A variant of LIGA is to replace the expensive synchrotron X-ray exposure step by excimer laser irradiation. Excimer lasers are rare gas halide pulsed lasers with a typical pulse length of 10-15 ns, capable of providing several joules/cm² into a beam about 1cm² in area and at discrete wavelengths depending upon the gas used (eg 248 nm with Krypton Fluoride, 193 nm with Argon Fluoride). At wavelengths of 250 nm or below, organic materials such as resist polymers are ablated and very high quality, deep images exposed. Typically each laser pulse will ablate a complex, submicron pattern to a depth of 0.1 - 0.2 μm over an area of 1 cm².

An excimer laser micromachining system projects a very high irradiance image (eg 100 MW cm⁻²) of a chrome-on-quartz photomask on to a resist and ablates the pattern directly ie. no chemical development step is required. A recent result is shown in Fig 8. The optical system can project an array of such images and at a pulse frequency of 100 Hz, produce such a pattern in approximately 10 sec. The process can be step-and-repeated in a manner similar to semiconductor manufacture and very large areas may be structured. For some materials, eg mylar and polyimide, laser ablation can "machine" the final product, eg microturbine rotors, directly as a one-step microfabrication process. The laser technique has the potential to make large numbers of microcomponents cheaply and to a good quality.

Thus, an elegant replacement for X-ray exposure is a standard chrome-on-quartz photomask, a laser illumination system and a deep uv lens. A microturbine manufactured by the process is shown in Fig 7, but with an excimer laser exposure step, is shown in Fig 9.

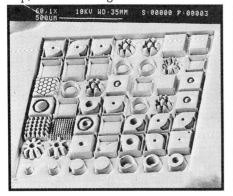 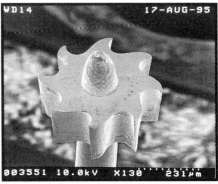

FIG 8 100-150μm deep resist structures ablated using excimer laser projection of a photomask. The laser wavelength was 248nm (KrF)

FIG 9 A nickel microturbine manufactured by electrodeposition in a mould similar to Fig 8

4.   CONCLUSIONS

The manufacturing techniques described in this paper originate from those already employed by the semiconductor industry to produce microelectronic and optoelectronic circuits. Variants of these techniques can now produce micromechanical parts of high precision, high complexity and high reliability. This presents the microengineer with a revolutionary tool-kit to manufacture novel microsystems of unprecedented specifications and performance.

anoparticles for drug delivery

ofessor S S DAVIS DSc
·d Trent Professor of Pharmacy, Department of Pharmaceutical Sciences, University of Nottingham, UK and
·nBioSyst (UK) Limited, Albert Einstein Centre, Highfields Science Park, Nottingham, UK

Synopsis

Nanoparticles can have an important role in drug delivery. By judicious choice of size and surface character it is possible to "direct" nanoparticles to tissue and cell targets following their injection into the body.

1. DRUG DELIVERY

Today, we have available a wide armamentarium of therapeutic materials that can be used as drugs or as diagnostic or prophylactic agents (e.g. vaccines). Such materials can take the form of relatively simple organic chemicals such as aspirin and penicillin while others are far more complex in their structure and can be carbohydrates such as the anti-blood clotting material heparin or proteins such as insulin and growth hormone. Diagnostic and vaccine agents can also display a range of complexity. The latter can be whole organisms such as (inactivated) viruses or the simpler surface or core structures of organisms that cause infection.

The advent of genomic science and the treatment of diseases using genes as drugs (gene therapy) heralds a new departure in medicine where the therapeutic agent is manufactured in the body. The administered 'drug' in the form of a nucleic acid (plasmid DNA), is huge in comparison to more conventional drugs (indeed, when in a compacted form plasmid DNA is a nanoparticle, but more of that later).

Some drugs are designed to act throughout the body and be deliberately unselective in their distribution. The treatment of infections with antibiotics is a good example of this approach. However, many drugs only have a beneficial action at one specific site (or sites) in the body. Their presentation at irrelevant sites can be wasteful, but more importantly can lead to side effects and adverse reactions. The use of chemotherapeutic agents in cancer treatment is a well known example. The drugs are used to kill rapidly dividing cancer cells, but can also cause damage to normal cells. Thus, in many cases it would be greatly advantageous to be able to deliver drugs selectively to their site of action in the body. This is not a new concept but owes its origins to Paul Erlich and his concept of the 'magic bullet'. Similar arguments can be put forward for other therapeutic modalities such as diagnostic agents and vaccines where a localized "effect" is required at a special site in the body for best advantage.

2. SITE SPECIFIC DRUG TARGETING

Many different approaches have been used to achieve site specific drug delivery (sometimes referred to as 'drug targeting'). These include inactive chemical "Trojan horses" known as prodrugs, that are converted to an active form by enzymes that reside specifically in certain tissues. Monoclonal antibodies and soluble polymers have also been considered for drug targeting. Each system has its advantages and disadvantages depending upon the nature of the 'drug' to be delivered and above all, the nature of the target site. Some tissues (and cells within such tissues) are more readily accessible than others and the therapeutic task may be readily accomplished by delivery of the drug to an appropriate organ in the body (first order targeting). However, more usually, the task is more sophisticated. The target is more specific, in the form of a special cell type or pathological lesion in the organ (second order targeting) or could even be a structure in a cell in an organ (third order targeting).

3. NANOPARTICLES

It is now being appreciated that in many instances when attempting such targeting activities, nanoparticles can be highly beneficial in their ability to provide site specific delivery and drug targeting. Moreover, a nanoparticle may be a suitable means to amplify a response through its particulate nature. For example, the surface presentation of antigenic material to the immune system using a nanoparticle can give rise to an enhanced immune response.

In our studies, nanoparticles are usually defined as colloidal particles less than 1 micron in size and can include both solid and liquid particles (i.e. microspheres, emulsions) as well as vesicular structures known as liposomes. As will be discussed below, the fate of such particles in the body, and consequently their use in drug delivery, is largely dictated by colloidal characteristics such as particle size and surface properties. Drug loading and release is more largely determined by the physical nature of the particle 'core'.

The properties of a nanoparticle loaded with a drug can be controlled by formulation techniques. The size, and particularly the surface of the particle, can be chosen with due consideration to the intended use of the system. Drug loading, into or onto the particle can be problematical, especially when a high 'pay-load' of drug is needed since the large surface area of the system provides an excellent opportunity for rapid release *in vitro* and *in vivo*. A drug that is too well entrapped in the particle may be released too slowly once it reaches the target site. Interestingly in some cases, the drug substance itself can comprise the core of the nanoparticle. Very insoluble drugs can be produced in nanoparticle size by appropriate processing (e.g. milling, controlled precipitation using supercritical fluids). In the case of gene therapy, compacted plasmid DNA is in reality a nanoparticle of 50-200 nm in size. The exact magnitude will depend on the size of the original plasmid and the efficiency of the compaction process. Compaction can be achieved using concepts of polyelectrolyte interaction (i.e. polyanionic DNA interacting stoichiometrically (or non-stoichiometrically) with cationic lipids or cationic polyelectrolytes).

3.1 Fate of Injected Nanoparticles

The importance of nanoparticle properties (particle engineering as we call it) can be well illustrated by considering the fate of a particle that is injected into the bloodstream (vascular system) (Figure 1). Where it finally ends up (to do good or harm) will be determined largely by its size and surface character. By having control over such factors it is possible to determine the design specifications that will dictate the required performance of a nanoparticulate system so that it can be delivered to a specific target (or for that matter to be retained for a long period in the bloodstream). Not surprisingly, some targets are easier to reach from the bloodstream than others. Targeting ability can often depend on how well the particle avoids capture by inappropriate sites as well as those properties of the intended target site that will allow recognition and uptake (or particle escape to extra vascular tissues).

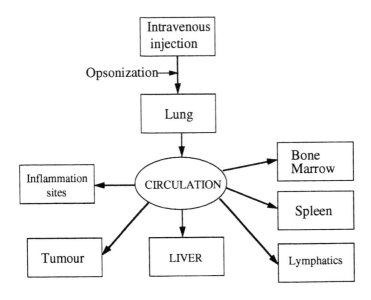

Figure 1

The fate of a nanoparticulate system after intravenous (i.v.) injection

As soon as a particle is injected into the blood it will be exposed to a highly complex environment. The blood contains a vast array of proteins and lipoproteins, antibodies and enzymes. These can be adsorbed rapidly to the particle to 'condition' or opsonize the surface. (Enzymes such as nucleases can degrade sensitive materials such as improperly compacted DNA). The quality and quantity of the adsorbed blood components will be controlled by the nature of the particle surface. Hydrophobic surfaces will adsorb more protein than more hydrophilic surfaces. We have found that it is possible to control the nature of the adsorbed components so as to have a direct influence on the fate of the injected particle. In general terms, a particle that is well coated with plasma components will be recognised as foreign and as a consequence removed rapidly by the defense systems of the body. This is the way in which invading microorganisms are cleared from the circulation.

3.2 Lung Uptake

Very large particles (10 μm) and above will be removed by a simple process of mechanical filtration in the capillary beds of the lungs. Such particles will therefore not reach the general circulation. This passive uptake process can sometimes be used for drug delivery when the target organ contains a suitable capillary bed. When using this method for lung targeting it will be important to adjust the dose of particles so that only a small fraction of the capillary beds are blocked by each dose (e.g. less than 10%) and that the carrier particles are able to degrade once the drug (or DNA) has been transferred into or across the endothelial cells of the blood vessel wall.

3.3 Liver Uptake and Kupffer Cells

Small particles (i.e. nanoparticles less than 1 μm) will be able to pass unhindered through the lung and thence into the general circulation. Particles that have been coated so that they are designated "foreign" will be captured rapidly and effectively by the liver. This capture is achieved by macrophage cells of the reticuloendothelial system called Kupffer cells. These cells reside in the blood vessels of the liver and act as scavengers. The nanoparticles are taken up by a process of phagocytosis; in a manner similar to the uptake of food particles by amoeba. In some instances the rate of uptake of nanoparticles by Kupffer cells can be controlled by the particle size provided the particles are not highly hydrophobic since these will be very effectively coated with recognition factors. (The Kupffer cell is endowed with a wide variety of recognition sites). In some disease conditions, such as infections by bacteria and certain parasites, the Kupffer cells of the liver are the therapeutic target. Consequently nanoparticulate systems loaded with drug can be highly effective at providing a targeting drug effect. However, in most diseases, the Kupffer cells represent an obstacle to delivery (and unwanted loss of drug). The macrophage cells are often efficient at degrading foreign materials to include drugs, especially if they are peptides or proteins, of polynucleotide in nature. In addition, drugs that are inadvertently delivered to the Kupffer cells may not be degraded but could cause damage to the cells, thereby affecting the normal defenses of the body.

3.4 Coated Nanoparticles

The avoidance of unwanted particle uptake by the liver has been a major task in the area of nanoparticle targeting. Our group in Nottingham were the first to do this using concepts of steric stabilization, familiar to those in the field of Colloid science. We reasoned that it should be possible to minimize the uptake of conditioning plasma proteins and to reduce particle cell interactions, by endowing the nanoparticle with a hydrophilic surface layer of sufficient dimensions as to provide a repulsive barrier (entropic in nature). We chose block co-polymers based on the hydrophilic material, polyoxyethylene (PEG). The PEG material was attached to the particle by a process of physical adsorption (the copolymer also had a hydrophobic block of polyoxypropylene that gave good attachment to the particle surface) (Figure 2). More recently, we produced particles where the PEG chains have been grafted to the surface or where the PEG is exposed at the surface through a process of self-assembly. We soon became aware that the efficiency of a PEG coating would depend on the surface density and the chain length of the PEG. The best effect was obtained when the PEG chains were extended into the external medium in the form of a 'brush-like' layer. Subsequently, we have found that a surface layer thickness of about 10 nm for the adsorbed hydrophilic polymer was effective (this corresponds to a molecular weight of PEG between about 2000-5000). The surface layer thickness will be dependent on various factors.

Nanoparticle coated
with block co-polymer

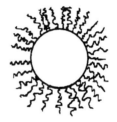

Nanoparticle grafted with
polyoxyethylene groups

Figure 2

Coated nanoparticles

3.5 Long Circulating Particles

Nanoparticles well covered by PEG (i.e. adsorption occurring in the plateau region of the adsorption isotherm) are no longer recognised by the Kupffer cells of the liver and can circulate in the blood for extended periods of time in contrast to the minimal circulation time for the uncoated particles. Data obtained with labelled polystyrene nanoparticles are shown in Figure 3 in the form of gamma camera images and in Figure 4 as histogram plots of liver and blood distribution.

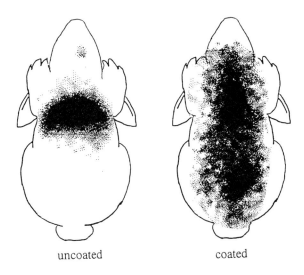

uncoated   coated

Figure 3

Gamma camera images demonstrating the whole body distribution of nanoparticles in the rabbit following intravenous administration. The nanoparticles, 60 nm in size, were surface labelled with Iodine-131. The images were obtained 7 h after injection.

The quantity of surface uptake can be controlled by adsorbing the block copolymer at different equilibrium concentrations that correspond to different regions of the adsorption isotherm. When this is done there is an excellent correlation between liver avoidance and polymer adsorption up to a point where the surface coverage by PEG is about 25% of its maximum value (plateau region). The extent of liver uptake therefore, can be controlled by a proper understanding of the underlying physicochemical processes.

Long circulating nanoparticles based on this concept have now been developed for drug therapy. The nanoparticles can be coated microspheres or emulsions prepared using self-assembling polymers (e.g. polylactide polyethylene glycol) or in the form of liposomes. A liposome product (known as a "Stealth liposomes") has been marketed. Herein, the polyethylene glycol is presented at the surface of the liposome as a conjugate with a phospholipid (phosphatidyl-ethanolamine). These liposomes can carry anticancer agents (doxorubicin) and have been approved for the treatment of Karposi's sarcoma and other conditions. Apparently these long circulating liposomes are able to escape from the blood to reach cancer cells in regions of pathology.

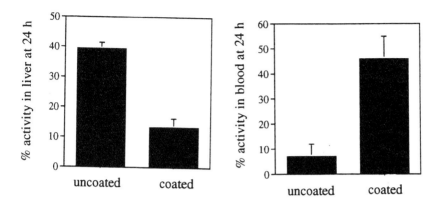

Figure 4

Distribution of coated and uncoated nanoparticles in the rabbit at 24 h following intravenous administration. The core particles were approximately 200 nm in size and prepared from a degradable polymer, polylactide-coglycolide. The particles were coated with a hydrophilic block copolymer, polylactide-PEG. Data for liver (left) and blood (right).

3.6 Delivery to Endothelial Cells

Nanoparticles, that avoid the macrophage cells in the liver, can be directed to specific target sites through surface manipulation or the attachment of specific groups (or ligands) that confer tissue or cell specificity. For example, in our own studies we have found that nanoparticles coated with a block copolymer of a certain molecular weight could result, not only in the expected effect that the particles were ignored by the Kupffer cells, but the surprising observation that the particles were captured exclusively in the bone marrow by endothelial cells that lined the blood vessels. The mechanism responsible is still to be elucidated. One hypothesis is that the specific block copolymer surface on the nanoparticle encourages the uptake of a blood component that provides the recognition effect. It is also appreciated that the flow of blood through the bone marrow blood vessels (sinusoids) is relatively slow, thereby giving a greater opportunity for particle recognition and capture.

3.7 Attached Ligands

Specific ligands, that can be used to promote particle delivery to designated cell types include monoclonal antibodies and sugar groups such as fucose, galactose and mannose. Target cells can have receptors for sugars that are presented in an appropriate conformational state. Particle targeting to the parenchymal cells of the liver is an example where a combination of particle size, surface and ligand mediated delivery is required. The parenchymal cells (hepatocytes) are the cells that make up the greater part of the liver and have a key role in metabolism. They can be important sites for drug targeting such as for antiviral agents in hepatitis or for gene therapy, where the parenchymal cells can be used as cellular factories to produce hormones, etc. The parenchymal cells are not as readily accessible as the Kupffer cells and endothelial cells. They lie outside the circulation system but can be reached by macromolecules and small nanoparticles. The endothelial cells lining the blood vessels of the liver do not form a continuous barrier but contain holes or fenestrations. (This is unlike other normal endothelial cell layers elsewhere in the body). These holes are of a maximum size of about 150 nm. Thus, in order for a nanoparticle to escape from the blood to reach the liver parenchyma, the particles need to be less than 150 nm in size and also to have a hydrophilic surface coating that prevents their capture by the Kupffer cells. Access to the parenchymal cells, however, does not necessarily lead to particle uptake. A receptor mediated process needs to be stimulated. The liver parenchymal cells have receptors for various ligands. A strategy suitable for drug targeting is to use the galactose receptors present on hepatocytes. These receptors are particularly sensitive to the presentation of three galactose groups arranged in a so-called 'triantennary' structure.

3.8 Targeting To Sites of Inflammation And Tumours

Other potential targets for nanoparticles that can be reached from the blood, include sites of inflammation and certain tumours. In these cases the pathological condition results in blood vessels being damaged and leaky so that nanoparticles have the chance to escape from the circulation either as single particles or carried across in migrating white blood cells.

3.9 Delivery To The Spleen

One further example can be described, where particle engineering can be used to good advantage to deliver drugs to specific organs via the blood circulation. The spleen has an important role in the immune system. It is also the place where senescent blood cells are removed by a process of filtration. (Old blood cells gradually loose their flexibility and can no longer squeeze through in the filtration 'beds' presented by the spleen). Hydrophilic nanoparticles and liposomes less than about 250 nm in size can pass unhindered through the spleen. However, if the particles are larger they will tend to become trapped by the filtration mechanism and later presented to the resident macrophages of the spleen.

3.10 Other Routes For Injectable Nanoparticles

Similar strategies based on particle engineering for the purposes of site specific drug delivery have been developed for other routes into the body such as following intramuscular or subcutaneous injection. For these two routes, nanoparticles will move from the injection site and enter the lymphatic system. This system is a second circulatory system of the body and has a key role in the transport of plasma proteins, particulates and dietary lipids. The lymphatics contain lymph nodes at intervals along their length. These nodes have an important function in the immune system and can be involved in the spread of cancer (metastasis) and HIV. For the latter condition it is believed that HIV can hide away in the lymph nodes for extended periods of time. The lymph nodes have a filtration capacity and also contain macrophages. Advantage could be gained if drug loaded nanoparticles or diagnostic agents are delivered selectively to regional lymph nodes. In our studies we have demonstrated that small nanoparticles (less than 100 nm) coated with polymers can result in high levels of targeting to lymph nodes in animals. More especially we found that the resultant particles needed to have intermediate properties in terms of their surface hydrophobicity/hydrophilicity. Particles that were too hydrophobic failed to leave the injection site and enter the lymphatics. On the other hand, particles that were too hydrophilic left the injection site, travelled in the lymph but passed unhindered through the regional nodes to eventually reach the systemic circulation (via the thoracic duct). Particles with intermediate hydrophilicity were discovered to give conditions for good movement from the injection site to lymph <u>and</u> recognition and capture by the lymph node macrophages (Figure 5). Such particles are now being investigated for drug delivery and diagnostic applications.

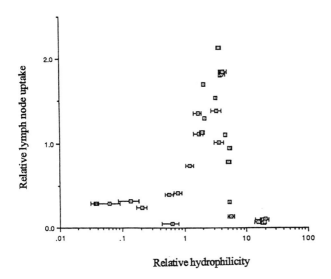

Figure 5

Uptake of radiolabelled nanoparticles in the regional lymph nodes following sub-cutaneous administration to the rat (foot-pad). Data were obtained 24 h after dosing. The particles, 60 nm in size, were coated with different block copolymers to alter the surface hydrophobicity/hydrophilicity. Relative hydrophilicity was measured by hydrophobic interaction chromatography. Relative lymph node uptake is calculated as the ratio of the quantity of particles in the regional nodes to the quantity remaining at the injection site.

4. REFERENCES FOR FURTHER READING

1   ILLUM, L., DAVIS, S.S., MULLER, R.H., MAK, E. and WEST, P. "The organ distribution and circulation time of intravenously injected carriers sterically stabilised with a block copolymer poloxamine 908". Life Sciences <u>40</u> 367-374 (1987).

2   ILLUM, L. and DAVIS, S.S. "Targeting of colloidal particles to the bone marrow". Life Sciences <u>40</u> 1553-1560 (1987).

3   ILLUM, L., WRIGHT, J. and DAVIS, S.S. Targeting of microspheres to sites of inflammation. Int. J. Pharm., <u>52</u> 221-224 (1989).

4   YOUNG, J.J., PIMM, M.V. and DAVIS, S.S. Targeting of a colloidal drug carrier using F(ab)2 fragments of monoclonal antibody. J. Pharm. Pharmac. Suppl. 40 133P (1988).

5   MOGHIMI, S.M., PORTER, C.J.H., MUIR, I.S., ILLUM, L. and DAVIS, S.S. Non-phagocytic uptake of intravenously injected microspheres in rat spleen: Influence of particle size and hydrophilic coating. Biochem. Biophys. Res. Comm. 177 861-866 (1991).

6   PORTER, C.J., DAVIES, M.C., DAVIS, S.S. and ILLUM, L. Microparticulate systems for site-specific therapy - bone marrow targeting. In Polymer Site Specific Pharmacotherapy. Editor A.J. Domb, Wiley, Chichester, p.157-203 (1994).

7   STOLNIK, S., DUNN, S.E., GARNETT, M.C., DAVIES, M.C., COOMBES, A.G.A., ILLUM, L. and DAVIS, S.S., TAYLOR, D.C., IRVING, M.P., PURKISS, S.C., TADROS, T.F., Surface modification of poly(lactide-co-glycolide) nanoparticles by novel biodegradable poly(lactide)-poly(ethylene glycol) copolymers. Pharm. Res. 11 1800-1808 (1994).

8   STOLNIK, S., ILLUM, L.I. and DAVIS, S.S. Long circulating microparticulate drug carriers. Adv. Drug Del. Rev. 16 195-214 (1995).

9   HAWLEY, A.E., DAVIS, S.S, and ILLUM, L.I. Targeting of colloids to lymph nodes: influence of lymphatic physiology and colloidal characteristics. Adv. Drug Del. Rev. 17 129-148 (1995).

POWER

e nuclear option

R HAWLEY DSc, FIMechE, FRSE, FEng, FIEE, FInstP, FInstNucE
ef Executive, British Energy plc, London, UK

SYNOPSIS

The world's consumption of primary energy continues to rise rapidly, mainly because of the developing countries who cannot yet provide the services essential to improving quality of life. Increasing energy consumption, the effect it will have on the world's finite resources and, more importantly, on the environment, leave the world's population facing serious challenges.

The paper will briefly consider the power generation technology options that offer sustainable development including the role that nuclear power plays today, and will need to play in the next century, to preserve and improve quality of life worldwide.

1. INTRODUCTION

The world, its people and their Governments, today face serious challenges in respect of energy usage and the resultant effects on the environment. These challenges lead to choices that will have far reaching consequences for the stability and even long term viability of life on our planet.

Energy gives us the ability to improve our quality of life. It provides the means for clean water and sanitation, essential for improving health at the most basic level. It provides power to our hospitals, giving us the facilities to detect illnesses and treat them effectively, prolonging our lives. It provides light, which enhances our ability to read and study, improving our educational capabilities. It provides heat for cooking, and power for refrigeration, allowing us to preserve the quality of food, enhancing health and nutrition. It provides warmth and cooling to increase our personal comfort, protecting us from environmental extremes. It allows our economies to grow through industrialisation, global communications and improved transportation.

In the developed world we have become accustomed to energy being available as and when we need it, for example, at a petrol station or by the flick of a switch. We take energy for granted and we do not consider what life would be like without it. Yet for half the world's population the amenity of a single light bulb in the home is not even an option (1).

But equally, the exploitation of energy can cause degradation in our quality of life, particularly when it challenges the safety and quality of our environment.

2. THE ENVIRONMENTAL CHALLENGES

Today environmental issues, such as acid rain, global warming, the depletion of the ozone layer, and the wider picture of global sustainable development, are firmly on the agenda.

2.1 Sulphur emissions.

Sulphur dioxide, one of the major contributors to acid rain results from the burning of coal and oil. It causes the acidification of lakes and rivers which in extreme cases can cause the death of fish and plant life.

Developed countries are taking steps to address the problem. Under the United Nations Economic Commission for Europe's second sulphur protocol (2), Britain and other European Countries have agreed to reduce sulphur emissions by 50% of their 1980 levels by the year 2000, 70% by 2005 and 80% by 2010.

Progress is being made. For instance, in March 1995, DTI statistics (2) showed that Britain would probably meet these commitments because of:
a)   reduced coal and oil burn,
b)   the introduction of Flue Gas Desulphurisation equipment,
c)   further commissioning of new Combined Cycle Gas Turbines in the electricity
         supply industry, and
d)   improvements in performance of nuclear power stations.

Other countries are also making good progress. Sulphur dioxide emissions from the largest electricity utilities in the USA fell more than 50% between 1980 and 1995 according to US Environmental Protection Agency (3) and many other OECD countries have reduced their emissions over the last two decades.

2.2 Global warming.

Greenhouse gases in the atmosphere, one of which is carbon dioxide, make life possible by letting in sunlight whilst preventing too much of the warmth created escaping back into outer space. Increasing concentrations of greenhouse gases in the atmosphere, due to human activities, such as burning fossil fuels and deforestation, appears to upset the balance and lead to a rise in global temperature.

It has been forecast that our planet could face a climatic shock unlike anything experienced in the last 10,000 years (4). But is this hysteria?

The Intergovernmental Panel on Climate Change (5) forecast in 1992 that worldwide carbon dioxide emissions could rise to over 70,000 million tonnes annually by the end of the next century, almost three times their current level. They predict that the resulting change in global temperature could be as much as 4°C. As a result the location of food sources, upon which many are dependent, could become disrupted. A warmer planet could also lead to thermal expansion of the oceans resulting in rising sea levels which would threaten many of the world's major urban areas situated in coastal regions.

Globally 1995 was the warmest year since records began in 1860. All but one of the warmest 11 years on record have occurred since 1980. Recently, a senior scientist from the US National Climate Data Centre stated that severe storms and flooding experienced by the US in March this year, were the type of events expected to become more frequent as global temperatures increase.

Experts and some Governments are convinced that the threat is real to the extent that, at the Earth Summit held in Rio in 1992 (6), a number of principles to support sustainable development were put forward. So far, 24 OECD countries, 11 economies in transition and the EEC as a region have accepted the aim of returning their Greenhouse Gas (GHG) emissions to 1990 levels by the year 2000.

In 1996, the UK government reported (3) that it expected to improve on this target by between 4% and 8%. However, in 1995, energy related carbon dioxide emissions by all OECD countries were around 4% higher than the 1990 levels (3). Projections are that between 11 and 13 of the OECD countries, who have accepted the aim of reducing GHG emissions to 1990 levels, could exceed their year 2000 targets by over 10%.

At its most basic level we could control our global greenhouse emissions by controlling our demand for energy, but the increasing demand, particularly in developing countries, is likely to make this option impossible. Indeed it will be exacerbated by ever increasing world population and urbanisation.

2.3 Population growth and urban development.

Today there are over 5.7 billion people on the planet and this is growing by over 90 million people each year (7), a number equivalent to about ten cities the size of London. With this growth rate, it is not surprising that the United Nations (7) estimates that the earth's population will be 8.5 billion by the year 2025 and 10 billion by the year 2050.

History has shown that, with a growing population, urbanisation occurs at a rapid rate. Global urban population has reached three times its 1950 level (1) and in the Third World the number of city dwellers has grown more than twice as fast as in industrialised countries (8). Urbanisation leads to increased economic activity and industrial development, bringing the prospect of better standards of living and improved quality of life for greater proportions of the population. However, the downside is that this leads to increased energy consumption resulting in increased airborne pollution and other environmental damage.

Earlier this year, the British Government Panel on Sustainable Development noted that most of the measures required to improve UK air quality affect transport. But the measures required to tackle the unsustainable growth in road transport, to improve public transport and to promote greener forms of transport less harmful to the environment are not yet in place. The panel recommended that stronger regulatory and fiscal measures are required if air quality is to be improved.

2.4 Energy demand

In the future, developing countries are likely to be the areas of major growth in energy usage, with rates at least 2 to 3 times greater than in developed countries (1). As a result, electricity and transportation will be the two fastest growing energy sectors.

The World Energy Council (WEC) addressed the growth and environmental issues in its 1993 analysis of world and regional energy needs over the next 30 - 100 years (1).

They concluded that, because of expected population growth alone, even with a modest increase in per capita energy consumption to no more than a subsistence level, there will be a global energy demand growth of between 30% and 95% by 2020. This will occur almost exclusively in the newly developing countries whose indigenous supplies are dominated by fossil fuels.

So it will be virtually impossible to limit man made CO_2 emissions to 1990 levels by 2020, even assuming the maximum contribution from conservation, energy efficiency and renewables.

Beyond the 2020 horizon, even if environmental concerns have not already driven down the use of fossil fuels, growing scarcity will increase dependence on alternatives sources of energy. Actions to encourage every aspect of reduced fossil fuel dependence to meet the needs of 2020 are essential. But what are the alternatives to fossil fuels for electricity generation?

3. OPTIONS IN ENERGY GENERATION TO MEET DEMAND

Developed countries have an important role to play in assisting developing nations to meet their energy needs with environment-led technologies. By increasing deployment of modern, efficient and clean technology, developing nations should be more able to meet their future energy demands in an environmentally sustainable manner, improving their quality of life and sustaining the global environment in the process.

However, ultimately, if there is to be a sustainable future, the world must move towards using more renewable energy, often cited as the only clean alternative to fossil fuels. Hydro, wind, solar, geothermal and wave power, as well as other renewables such as biomass and tidal energy, all have the potential to go part way to meeting the forecast increases in demand, whilst reducing global warming emissions, acid rain and urban air pollution.

One area where some renewables can play a major role is in remote, developing, sparsely populated locations where a local energy supply is required and where access to a major transmission network is too costly. However, in urbanised areas where the concentration of energy demand is high, the variability of many renewable sources are not suited to supplying base load electricity.

The World Energy Council concludes that extensive penetration of renewables into the energy market is not likely to happen quickly (9). There is a natural time lag between the success of prototypes and demonstrations leading to large scale commercial applications (9). But the length of time needed before they can make a significant contribution is unknown and the longer we wait the more critical our environmental situation may become.

However, there is another source of energy that is already available that will help secure sustainable urban and global environments and quality of life. It is available to provide large volume, baseload electricity if there is a will to use it. That source of energy is nuclear power.

4. THE NUCLEAR OPTION

4.1 Current situation

Nuclear power provides 7.3% of the world's total energy production (10) and meets around 17% of the world's electricity demand (11). At the end of 1996, 443 nuclear power reactors were in operation worldwide (11) and in 1995 nuclear power generated more than 2200 TWh of electricity. This is greater than total electricity production from all sources in 1958 when nuclear power was in its infancy. Overall nuclear power is the second most used source of electricity in Europe, providing just over one third of its electricity needs (11).

But nuclear trends are changing. The centres of rapid growth in nuclear capacity are shifting away from those parts of the world that pioneered nuclear technology. Instead interest in nuclear power is now growing fastest in the "Tiger" economies of East Asia and the Pacific Rim.

4.2 Environmental benefits

Nuclear power as an alternative to fossil fuels produces negligible emissions of sulphur oxides, nitrogen oxides and carbon dioxide (11).

In 1995/96 the UK's nuclear power stations saved the emission of 87.56 Mt. of carbon dioxide that would have been produced had the equivalent amount of electricity been generated from burning coal (11). This made a crucial contribution to meeting the UK's carbon dioxide emission target of reducing emissions to 1990 levels by the year 2000. Applying this type of statistic across the globe highlights the environmental benefits of nuclear power.

In addition, compared with coal-fired power stations relatively small quantities of fuel are required, so there is less environmental impact from mining, transportation and ash disposal.

However nuclear power is not without its problems.

4.3 Challenges to overcome

There are a number of challenges facing nuclear power, the size of which varies from country to country.

The most direct and serious challenge to the successful use of nuclear power, in providing the energy needs of future generations, is the ongoing problem of public perception and political will. They go together in that each has an influence on the other and politicians, in the end, have absolute control over the market environment through their treatment of regulation and subsidy policies.

Two decades of public and political debate in Sweden has recently resulted in a Government decision to close the country's 12 nuclear power plants. These power plants currently provide half of the country's electricity needs. However, viable alternatives are still to be developed and both public and industrialists alike have expressed concern at the decision, not least because of the rise in electricity prices that are likely to result. In addition, in the short term, Sweden will import electricity generated by fossil fuels in neighbouring countries in order to meet demand until renewable supplies are available - whenever that might be.

The Japanese Government are firmly committed to nuclear power which currently provides over a third of the country's electricity. However, there is growing public concern over safety following a sodium leak from a secondary (non-radioactive circuit) at the Monju prototype fast breeder reactor in December 1995 and two fires and an explosion at the nuclear fuel reprocessing plant at Tokaimura in March this year. Strength of public opinion was expressed in August last year when local residents participating in Japan's first nuclear referendum, rejected plans to construct a new nuclear power plant at Makimachi.

Add to this the recent public protests in Germany concerning waste storage and transportation, then it is clear that to gain greater global public acceptance and trust, the nuclear industry must continue to establish the right track record and provide good reasons why nuclear should be wanted, not just needed. High safety standards, proven solutions, openness and two-way communication are essential to the achievement of this objective.

Worldwide, nuclear operators aim to achieve the highest standards in safety and more efficient operation. The industry recognises that any breach of safety represents a serious threat to the industry as a whole.

With notable exceptions, it has been said that nuclear reactor designs have demonstrated they produce safe power, so their next overriding challenge is to produce cheaper electricity.

In a purely commercial situation, involvement in a nuclear power programme can only be justified, and indeed financed, if it provides electricity at a cost which is competitive with alternative fueled plant. Currently, in the UK, Europe and the US, the target generation cost for nuclear power is set by the cost of generation from new fossil fueled power plant. In the UK specifically, this target is currently set by Combined Cycle Gas Turbine (CCGT) plants, the latest generation of which should achieve around 60% efficiencies.

But because of the relatively low capital cost, the cost of electricity generation from CCGT is particularly sensitive to the cost of natural gas. Today the availability of cheap natural gas has resulted in the cost of generation from these units dropping to very low levels.

Thus, in Europe and the US, where there are abundant supplies of gas available at a relatively low price, CCGT plant have set a very demanding target for new nuclear plant.

However, it is conceivable that gas prices may rise, especially as resources become depleted. In such circumstances, new nuclear build would become more attractive.

4.4 Developments in nuclear power technology

The most recent addition to the nuclear power fleet in the UK is Sizewell B, owned by Nuclear Electric Ltd, part of the British Energy Group. It consists of a modified Westinghouse PWR design with a capacity of 1200 MW. On 22nd September 1995 the Station received it's rating certificate from the Nuclear Installations Inspectorate confirming full commercial operation. As such, it achieved commercial load very close to programme and within the budget sanctioned in 1990. Since then, the station has set some impressive records. For example, its first refueling outage, completed in a record time of 55 days, was the third shortest time for the first outage of any Westinghouse four-loop PWR plant and broke the world record for having the lowest radiation dose rates for a standard Westinghouse PWR.

Of the world's nuclear generating capacity in 1995, Pressurised Water Reactor (PWR) technology represented 63.9%, Boiling Water Reactor (BWR) technology 22.5% and Heavy Water Reactor technology 5.5% (12).

Out of the 40 reactors under construction at the end of 1996, 31 are light water reactors and 9 are pressurised heavy water reactors (13). It is likely, therefore, that proven water reactor technology will remain dominant in the medium term.

The latest development in water reactor technology is the European Pressurised Water Reactor which is a joint venture being developed by Framatome and Siemens. With a capacity of 1450MW, and a lifetime of around 60 years, the EPR is expected to be available by the end of the century.

In the US three designs are under development. The first of General Electric's 1350 MW Advanced Boiling Water Reactors achieved commercial operation last year in Japan. Its modular design and prefabrication construction techniques helped achieve a 4 year and 3 month completion time from inspection of bed rock to commercial operation. Further plants, similar to this design, are planned for construction in Japan and Taiwan. In the US, the design received Final Design Approval from the US Nuclear Regulatory Commission (NRC) in July 1994.

ABB Combustion Engineering's System 80+ is a 1350MW advanced PWR station which offers enhanced safety systems and simplification to reduce costs. It too achieved Final Design Approval from the NRC in the summer of 1994.

The Westinghouse AP600 is a move towards a smaller integrally safe design of reactor system. The attraction is a simpler design with increased passive safety systems that rely on gravity, natural circulation and stored energy for accident prevention and mitigation. At around 600 MWe, the AP600 represents a better match to the market in countries with a developing electricity infrastructure and the Chinese are expressing interest.

These three designs are essentially refinements, to a greater or lesser extent, of existing reactor technology. But what might the future bring in terms of nuclear technology?

4.5 Fast reactor

In most realistic energy growth models, fast reactors could be required around the middle of the 21st century, due to depletion of currently known Uranium deposits. Anyone who has visited the 'Superphénix' 1240 MW prototype Fast Reactor at Creys-Malville, near Lyon, will recognise a technology that is scientifically and technically proven at a commercial scale. At present, this type of plant is not considered able to compete with current light water reactor technology due to costs of construction. However, it is easy to imagine that it could compete, following limited further development and a commitment to a sufficiently large programme, to bring the full benefits of scale already demonstrated by the French in their PWR programme. A commercial fast reactor is feasible today.

However, if the fast reactor is to be a major part of our energy future, it must be developed further to ensure that the commercial risks are clearly understood and minimised well before there is a major commitment. This cannot be achieved without further investment.

Currently, the French and Japanese are most likely to undertake such a significant investment and both their fast reactor programmes are suffering from setbacks. The French Superphénix reactor's 1994 restart licence has been annulled and there are calls for an inquiry into the future of the project. The Japanese Monju reactor remains shutdown following the sodium leak in 1995 and the date when it will return to power is uncertain. However, it appears that the Japanese Government remain committed to funding research and development of fast reactor technology.

4.6 Fusion for the future?

Fission and fusion were discovered within a few years of each other and, while fission power is now a maturing industry, fusion has yet to demonstrate scientific feasibility. There has been enormous progress in understanding the physics of tokamak fusion over the past 40 years, and yet this first stage is still incomplete. The science is proving exceptionally hard to master and the practical engineering will be much harder still.

There are two fundamental engineering issues. The first stems from the geometry of a tokamak. To remove the heat from the fusion reaction requires the handling of very high heat fluxes, some ten times greater than the maximum found in a modern PWR core, or the production of lower flux densities. This can only be achieved by significantly increasing the size of the fusion reactor.

The second stems from the nature of fusion energy, 80% of which appears in the form of fast neutrons which damage structural materials, degrading their mechanical properties. Even if new super-materials could be developed which out-perform today's, the first wall of the structure would probably have to be replaced every few years.

The next planned step in harnessing fusion is the International Thermonuclear Experimental Reactor (ITER), which will have a total project cost exceeding $10B. ITER partners are Japan, the European Union, Russia and the US, but at present no firm decision to build ITER has been taken and indeed there are ongoing debates over the feasibility of the proposed design.

Fusion is a research field with some promising lines and many cul-de-sacs. However, breakthroughs in science and technology can never be discounted as Robert Millikan winner of the Nobel Prize in Physics in 1923 found to his cost. He once said: "There is no likelihood that man can ever tap the power of the atom."

5. CONCLUSION

To achieve sustainable growth of the world's economies and their energy supplies, the world's population must become less dependent on fossil fuels. This is especially so, if this growth is to be achieved without increasing irreversible damage to the environment which would threaten our own and future generations' quality of life.

In theory, renewables can go some way to bridging the gap, but can they be relied upon, in the timescales required, to supply over 50% of the worlds' primary energy as implied in one of the WEC's scenarios?

By contrast, nuclear power is already technically capable of meeting the challenge. However, it needs to win the support of both Governments and the public. In some countries where market conditions favour 'short term' solutions to energy supply, there may need to be a strategic intervention by Governments to ensure that nuclear power has its place in a diverse energy mix.

In the meantime, it is vital that the nuclear power generating industry continues to operate safely through the adoption of best worldwide operational practice, promoting the use of inherently safe plant designs and the employment of quality people. In parallel, the industry must continue to reduce the cost of new build and operation.

6. REFERENCES

1   World Energy Council, 'Energy For Tomorrow's World', 1993.
2   Department of Trade and Industry, 'Energy Paper 65 - Energy Projections for the UK', March 1995.
3   World Energy Council, 'Climate Change Negotiations: COP-2 and Beyond', Report No.6, September 1996.
4   Mintzer, I.M.: 'Living in a warming world', Chapter 1 in Mintzer I.M. (Ed): 'Confronting Climate Change - Risks, Implications and Responses', 1992.
5   Intergovernmental Panel on Climate Change [IPCC], 'Climate Change 1992: supplementary report to the IPCC scientific assessment', 1992.
6   The United Nations Conference on Environment and Development (UNCED), 'Earth Summit '92', June 1992.
7   United Nations Population Fund [UNFPA], 'The State of World Population 1994'.
8   World Wide Fund For Nature, 'Atlas of the Environment', 1992.
9   World Energy Council, 'New Renewable Energy Resources, A Guide to the Future', 1994.
10   The British Petroleum Company plc, 'BP Statistical Review of World Energy 1996', 1996.
11   British Nuclear Industry Forum, 'Nuclear Trends', Issue No 2 - December 1996.
12   Kurtz D, 'The Future of Nuclear Power in Europe', 1996.
13   Nuclear Engineering International, 'World Nuclear Industry Handbook 1997', 1997.

Alternative energy sources

Professor N W BELLAMY
Pro-Vice-Chancellor, University of Coventry, UK

SYNOPSIS

Alternative energy research and development received a major boost during the 1973 oil crisis. Wind, wave, tidal, landfill waste, geothermal and other renewable energy sources were funded by government and industry in the expectation that they would be required to replace the declining fossil fuel sources. In the event the majority of these new energy sources could not compete with oil, gas and coal, particularly in the UK.

The paper covers the exciting developments in alternative energy and highlights the successes both in technical and commercial terms. Examples of the author's own research are presented to illustrate the challenge of the renewables.

1. INTRODUCTION

Energy in its natural form is well understood but is difficult to quantify. Everybody has felt the heat from the sun, observed the strength of the wind, heard the thunder of a waterfall and seen the waves crashing on to the shore. The amount of energy dissipated in these natural phenomena is substantial but has to be compared to that measured in man-made machinery. The heat given by a one bar electric fire (1kW) or the mechanical output of a small car (100kW) are every day measures of power and energy. A 1kW fire running for one hour consumes one kilowatt hour of electrical energy and costs around 6p at the domestic meter. This 1kWh is generated at a power station by about 3kg of coal or about half a gallon of fuel oil. The same 1kWh in the form of mechanical power can lift a human being from sea level to the top of the Matterhorn (4500m) in one hour.

Although renewable forms of alternative energy are abundant and available at most places on earth, it is by its very nature diffuse and difficult to harness for the benefit of man. Fossil fuels, at present so readily available, are very competitive and provide attractive cost-effective solutions to mankind's energy needs. They are a passing phase and within decades will have to be replaced by renewable sources or nuclear power in some form. The majority of people recognise this fact but only a few are prepared to face the consequences. The energy crises of 1973 and 1979 are still reminders of the importance of energy in every day life and highlight the need to be prepared for the ultimate decline in power derived from fossil fuels.

2. RENEWABLE ENERGIES

On a global scale the new renewables, as distinct from traditional biomass and large scale hydroelectric, account for only 2% of the global primary energy supply. Many energy scenarios have been predicted for the next 100 years during which the world population is expected to double. Biomass is generally expected to continue to dominate in the shorter term with a shift to solar in the longer term. The move towards sustainable energy use will be difficult to achieve in the likely time frame and will certainly depend on the development of the renewables.

2.1. Wind power

The renewable success story of the 1990s is the appearance of dozens of wind farms on the UK landscape. The environmental conflict between this green form of energy generation and the visual impact of the windmills themselves appears to have subsided. Early fears of local people at the planning phase about visual appearance, noise and TV reception have generally proved unfounded and in some cases beneficial such as tourist attractions. Clearly the majority of people believe the benefits of wind power outweigh the perceived disadvantages. The economics of wind power are looking more attractive and Europe has taken over the lead in installed capacity from the United States. European targets for renewable energy, 8% of the primary energy supply by 2005, suggests that the installed wind power capacity should be at least 10GW at that time. With the present rate of development this target should be achievable.

2.2. Solar power

Two methods are used to collect energy direct from sunlight, solar thermal and photovoltaic. Solar thermal devices convert sunlight into heat whereas in solar electric plant the heat is used to drive a heat engine coupled to an electrical generator. Photovoltaic devices convert solar photons into electrical charges that generate a current when connected to a load. Both technologies have great potential but the current economics are such that only niche markets are being served.

2.3. Biomass

The direct combustion of biomass to provide low grade heat is a major source of energy worldwide. In Europe many countries depend on heat energy derived from wood, peat, waste and refuse. The widespread use of bio-derived liquid fuels is not anticipated, as in some other countries, because they are regarded as prohibitive in cost. Without biomass a sustainable energy future would be impossible and positive energy policies are required to protect this vital energy source.

2.4. Geothermal

Natural geothermal is an important form of energy in some countries but although 'hot dry rocks' are only a few kilometres away from everybody this massive source of energy is proving difficult to harness. Apparently injection of water through boreholes into fractured rock reservoirs is only partially recovered and at a lower temperature than required for commercial exploitation.

2.6. Water and energy

Water is one of nature's gifts. Its link with solar energy is there for all to see. Sunshine can raise it from the oceans and sun-born winds can carry it to regions where it condenses to feed great rivers. The same winds interact with the surfaces of the oceans and create ocean waves. Both rivers and waves are sources of renewable energy on a global scale and have on their own to provide all mankind's needs. Hydroelectric power stations that extract the energy from rivers are well established source of electrical energy and provide more than one third of the demand even though only 12% of the potential has been exploited. Tidal energy has been demonstrated outside the UK but appears too expensive for the large Seven estuary site. Wave energy, on the other hand, is at an early stage of development end despite its enormous potential only experimental wave plant has been built. The long term benefits of these 'energy from water' technologies mean that no effort should be spared in their development and commercial exploitation.

3. HYDROPOWER

The harnessing of the energy from running water to provide mechanical or electrical power has been one of man's greatest achievements and yet is not related to any particular inventor. References to its development can be traced back over 2000 years. One might think that the development of hydroelectric power has run its course and this might be true except for the very low head range of applications. The market for small hydroelectric installations has increased dramatically during the 1990s. With the widespread concern about global warming and the growing support for renewables in many countries the price on offer for renewable energy such as hydroelectric power is proving very attractive. In the UK the privatisation of the generation and supply industry has led to the Non Fossil Fuel Obligation (NFFO) guaranteeing the market and price of electricity from hydro sources of around 4 to 6p/kWh as against 2.5p/kWh previously on offer. In addition institutional factors such as rates and water charges have been eased. Similar market changes are taking place in other countries and the economic scene is set for a period of considerable exploitation of many forms of renewable energy.

3.1 The Syfogen air entrainment system

To achieve respectable power outputs from low head sources the energy contained in large flows of low pressure water has to be extracted. If this low grade water power was converted into air power then high speed turbo-generators could produce the electrical power. In effect a water to air power converter would act as a transformer or gearbox and large expensive water turbines replaced by much smaller and relative low cost air turbines. There are a number of ways of converting water power to air power. Pneumatic box systems (1) have been proposed and tested whereby air is driven in and out of an enclosed chamber by water controlled by valves. High efficiencies can be achieved by such twin box arrangements but the valve mechanics are demanding in engineering and reliability terms. An alternative bag system (1) uses an automatic switching membrane between water and air but also has design problems as well as variable power over a 10 second cycle. In the event one of the simplest ways of devising a water to air converter is the entrainment of air into a syphon or venturi.

Both the venturi and the syphon can be used to reduce pressure in ducted water flow systems. Figure 1 shows the Syfogen (2) configuration that combines a venturi with a syphon to produce a practical low head hydro structure. The upstream bell mouth intake feeds the venturi throat at the syphon high point located over a river weir. Air is induced through injectors at the throat, where the pressure is the lowest, and entrained into the water flow in the mixer section. The two phase mixture of water and air then flows through a diffuser, to recover the velocity head, to the downstream exit where the air exhausts to atmosphere. To prevent separation of the air by buoyancy the duct section is profiled in an anti-gravity curve from injector to the exit bend. At the exit the reverse bend in the duct forces air towards the surface and directs the water downstream. The process converts a proportion of the water power into injector air power that is dissipated in turn by air flow through the air turbine. The air entrainment reduces the effectiveness of both the venturi and the syphon and two phase flow losses can be significant. Reasonable efficiencies of water to air power conversion can only be achieved by careful geometrical design of the Syfogen duct and by an air injection method which induces effective mixing. The characteristics of the system are such that high air resistance through the turbine reduces the injector air flow and increases the negative air pressure whilst low air resistance increases the air flow and lowers the pressure. Water flow varies inversely with air flow and the injection of too much air can disrupt the venturi flow and kill the syphon.

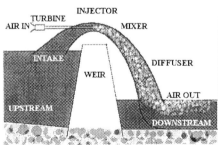

Figure 1. Configuration of the Syfogen   **Figure 2. Syfogen tests on River Derwent.**

Research work on hydroelectric power systems using pneumatic conversion has been carried out at Coventry University since 1986. Early experimental work on "box" systems was followed by demonstration tests of a full scale "bag" system at Borrowash on the River Derwent. The problems encountered with these two cyclical devices led to the Syfogen concept with its simple structure and constant power operation. A near full scale model of the

Syfogen was designed, manufactured and installed at Borrowash. The unit shown in Figure 2 was the largest size that could be accommodated at the restricted site and generally operated with a 2.3m head. In 1994 the performance of Syfogen demonstrated its high reliability and reached 22kW air power at 30% water to air power efficiency. With further development work it is predicted that a single unit at a 2.5m site would have an air power rating of 100 kW and an overall water to wire efficiency of 40%.

The Borrowash Syfogen models demonstrated that air entrainment was a feasible method of extracting energy from run of the river weir sites. The models proved to be 100% reliable and extremely easy to operate. Starting up and closing down a unit is by priming and unpriming the syphon intake. The only occasions the units stopped operating were when flood conditions reduced the available head to less than 1m. In normal operation the units were found to cope with changes in river levels, ice conditions, normal river trash and fish migration both downstream and upstream. Oxygenation of the water proved a particular environmental benefit to the river and fish. The advantages of this alternative hydroelectric power concept were obvious to all observers.

4. WAVE POWER

The Energy Systems Group at Coventry University has been researching wave energy and developing wave energy devices since 1975. With the help of Sea Energy Associates Limited and the UK Department of Energy the SEA Clam device was developed and tested in 1979 followed by the reconfigured version, the circular SEA Clam, in 1984. A series of model tests at Loch Nests proved the device concept and full scale designs in steel and concrete were refined in 1991. These designs were used in an in-depth review of wave energy by the UK Department of Energy who reported its findings in December 1992 (3). Five major designs were included plus reference to three more recent concepts. The report summarised its findings by comparing the major devices regarding their state of development and the predicted electricity generating costs. It concluded that wave energy is at a relatively immature stage of development and that the shoreline OWC had the lowest generating costs followed closely by the offshore SEA Clam (4). This cost comparison has to be seen in the light of the potential lower resource exploitable by offshore devices as against onshore units.

4.1. The oscillating water column

Wave energy research in Europe is concentrating on developing oscillating water column (OWC) devices. Examples are the OWC on a cliff face in the Azores, the Osprey OWC on the sea bed at Dounreay and the shoreline OWC on Islay. The Islay unit shown in Figure 3 sits over a natural shoreline gully that concentrates waves and reduces construction costs. The sea water oscillating in the gully is used to pump air through a turbine coupled to a generator and thus produce electricity. The unit has been operating since 1991 and has demonstrated the feasibility of wave energy as well as providing useful research data.

4.2. The Circular SEA Clam

The SEA Clam, shown under test on Loch Ness in Figure 4, is a simple device which uses the displacement of air to extract energy from sea waves. Twelve air chambers , the outer surfaces of which are formed by flexible rubber membranes, are placed around a floating ring structure.

Differential wave action moves the membrane in and out forcing air to be interchanged between chambers. Wells turbines placed in the manifolds between the air chambers extract power from the air flow. The rigid torus structure, 60m diameter or more, acts as a stable reference body and is moored a few kilometres off shore. Typically a 25MW scheme deployed off the West coast of Scotland would feature 10 SEA Clam units and produce over 50 GWh per year of electricity.

Figure 3. OWC installation on Islay.   **Figure 4. Model tests of Sea Clam.**

The SEA Clam has been designed to operate in the energetic seas off the west coast of Scotland where average power levels can reach 70 kW/m and with storm power levels of over 3000 kW/m. Predicted annual capture efficiencies for a 60m diameter SEA Clam in the characteristic sea states off South Uist, Scotland, were calculated as 33% in the wave energy review. This gives an average power per device of 600kW, an energy output of 5GWh/year and an electricity cost of 8p/kWh. The structural design is governed by the 25 year fatigue levels induced by these wave climates and the power system design relates to average sea power at a given site. Costs included capital costs of construction, installation, transmission and project management plus the operating and maintenance costs..

5. CONCLUSIONS

Renewable forms of alternative energy are important to the health of the world, need more attention and give rise to challenging research projects.

6. REFERENCES

(1) Bellamy, N.W. : Low head hydroelectric power using pneumatic conversion, IEE Power Engineering Journal, No3, pp109-113, May 1989
(2) Bellamy, N.W. : The Syfogen low head pneumatic hydroelectric system, Hydropower into the next Century, Barcelona; June 1995.
(3) Thorpe, T.W. : A Review of Wave Energy, ETSU Report for the Department of Trade and Industry,UK, ETSU-R 72, December 1992.
{4} Bellamy, N.W. :High efficiency, low cost, wave energy conversion system, ODEC International Conference, Muroran, Japan; 1992.

as turbine: combined heat and power

; **KENNEDY** BSc, AMIMechE
ctor, Power Business, Balfour Beatty International Limited, Sidcup, UK
HALL BSc, PhD, MIMechE, MAmSAE
cipal Engineer, Balfour Beatty International Limited, Sidcup, UK

SYNOPSIS

Combined Heat and Power (CHP) otherwise called Cogeneration, is the integrated production of electricity and useful heat. CHP has been used in industry since the last century, but its profile has increased over the past ten years, with a steady growth in the number of plants installed. New capacity has been dominated by natural gas fuelled gas turbine based plants.

This paper provides a simple introduction to the thermodynamic principles behind gas turbine CHP, outlines the history of its development in relation to pure power generation, and describes some associated practical aspects of engineering and economics, together with the benefits to the quality of life.

1. INTRODUCTION

Combined Heat and Power (CHP) otherwise referred to as Cogeneration, is the integrated production of electricity and useful heat. This straightforward definition places no constraints on the scale or nature of the plant, nor on the fuels employed.

CHP has been used in industry since the last century. Its profile has, however, increased over the past ten years, with a steady growth in the number of new CHP plants installed in industrial, commercial and domestic premises, the former typically based around gas turbine prime movers.

This paper aims to give a broad view of the current practice and potential future of CHP, its position within the energy industry as a whole, its sensitivity to public policy and its potential benefits to the quality of life.

Since this paper is prepared as part of the IMechE's 150th Anniversary Symposium and is thus aimed at a broad audience, a simple introduction to the engineering principles that form the basis of CHP may be helpful.

2. THE ENGINEERING BASIS FOR CHP

2.1 Thermodynamic fundamentals

The fundamental laws of thermodynamics, as applied to heat engines, give us the following axiom:

> "For a machine to produce work, it must receive heat from a source and it must reject a proportion of that heat to a sink at a lower temperature."

This is illustrated in figure 1.

For given temperatures of the source and sink (T_1 and T_2 respectively), the maximum theoretical efficiency of work (power) generation is the so-called Carnot efficiency:

$$\eta = \frac{T_1 - T_2}{T_1} \qquad \text{(Where T is in Kelvin)}$$

In practice such theoretical efficiencies cannot be attained and this simple relationship does not apply directly to practical power generation plant cycles. Nevertheless this fundamental equation dictates that an efficient power generation cycle requires heat to be input at as high a temperature (T_1) as possible, and rejected at as low a temperature (T_2) as possible.

One of the major constraints to the development of better efficiencies in heat engines is the choice of working fluid. Air or steam is inevitably chosen - the choice is pragmatic - these are the fluids that surround us.

These fluids do, however, have major thermodynamic disadvantages. In a conventional steam turbine power plant the inlet temperature (which determines T_1) is typically around 540°C, this being a metallurgical "break point" above which the alloys required for the steam system become uneconomic. Whilst it is possible to achieve rejection temperatures (T_2) only slightly above ambient, a significant amount of energy (around 50%) is thrown away when the steam is condensed into water so that it can be pumped. Gas turbines which use air as the working fluid, can have significantly higher values for T_1, the practical maximum temperature being around 1100°C. Unfortunately, T_2 is also high (a typical exhaust temperature is 500°C), an inevitable result dictated by the physics of the compression, combustion and expansion processes. Simplified schematic, cycle and heat balance (Sankey) diagrams illustrating these points are shown in figures 2 & 3.

Neither process can achieve a wide range between T_1 and T_2 and thus the highest theoretical efficiency that could be achieved (Carnot efficiency) is around 60%. In practice the actual overall efficiency is little more than two-thirds of this, around 40% in both cases.

Whilst there have been experimental cycles offering much better theoretical efficiencies, the obvious practical difficulties of dealing with working fluids such as mercury have prevented these becoming a practical reality.

2.2 The development of more efficient generation plant

From the above it is apparent that in order to improve the efficiency of the power generation process either T_1 must be increased, T_2 reduced or alternative cycles developed. Major step changes in efficiency are difficult to achieve.

2.3 Steam plant

Almost all of the major gains in steam cycle efficiency have now probably been realised, although incremental improvements in metallurgy continue to allow higher steam temperatures, cycle optimisation remains a balance between capital cost and operating benefit for each application. The major step in reducing T_2 was taken very early in the development of steam power when condensing engines were introduced allowing T_2 to approach ambient temperature by placing the exhaust under vacuum. Developments during the 20th century have raised the maximum T_1 from steam engine levels below 200°C to the current limit of about 600°C (governed by metallurgical considerations).

Modern steam turbine power plants have an overall efficiency (net power generated ÷ net fuel energy input) of about 40%. It is perhaps also worth noting, however, that environmental considerations such as water conservation and cooling tower plume abatement are increasingly forcing the use of air cooled condensers with a consequent increase in T_2 and reduction in efficiency.

2.4 Gas turbine plant

The gas turbine has been the subject of experiments since the turn of the Century but was first developed as a practical device during the late 1930's. Originally intended for shaft power generation, its development was driven during the Second World War in the form of turbojet engines for aircraft propulsion - most famously by Whittle, whose first engine ran in 1937.

The gas turbine is normally designed as an open cycle where air is fed in, compressed, heated by combustion with fuel, expanded through a turbine producing power, and finally exhausted to atmosphere. The keys to high efficiency of the open cycle are higher pressure ratio and combustion temperature (the higher the latter, the higher the mean temperature T_1 at which heat is added). In principle a closed cycle, using higher pressures and working fluids such as Helium which are continuously recirculated and heated by an external source, will give a more compact and efficient plant. Historically, open cycle propulsion engines dominated and in power generation the increased cost, size and complexity of the closed cycle arrangement outweighed the efficiency advantage.

The simple open cycle gas turbine has become economically more viable. Improvements in aerodynamic design of the internal components, combustion control and the availability of high temperature materials have enabled better efficiencies. Current gas turbines derived from aircraft units and modified for mechanical drives or power generation attain efficiencies exceeding 40%, with typical maximum temperatures of 1100°C. Current materials development, could increase this beyond 1300°C. Unfortunately, as the exhaust temperature (T_2) will also increase, the improvement in open cycle efficiency is limited. The main enhancements to the basic cycle have been based on the following principles:

- Increasing T_1 - using additional combustors to reheat the combustion gases part way through the turbine, so that more heat is added at higher temperatures. This is an effective way of increasing cycle efficiency but ultimately T_1 remains limited to the maximum temperatures above.

- Reducing T_2 - using a heat exchanger (recuperator) the hot exhaust gas can heat the incoming combustion air, thus reducing the mean temperature at which the exhaust heat is rejected and improving efficiency. The potential increases are limited, but developments of this principle have attracted interest where the relative simplicity and compactness of the system are important, for example in marine propulsion.

2.5 Combined cycle plant

From the thermodynamic restrictions discussed above it is apparent that real gains in the overall efficiency of power generation can only be achieved through a different cycle.

Fortunately the steam turbine and gas turbine have inherently complementary characteristics. The steam turbine has a low T_2 and limitations on T_1 whilst the gas turbine has a relatively high T_1. Figure 4 shows the two cycles integrated into a CCGT plant. The high T_1 of the gas turbine cycle is combined with the low T_2 of the steam turbine cycle and a synergy is achieved giving a Carnot efficiency of almost 80%. Once again for the practical cycle and plant the actual overall efficiency is around two thirds of this or a maximum of 55% in current plant, with 60% as a goal. The combination of cycles has other benefits. In the steam turbine cycle T_1 can be reduced without any efficiency penalty resulting in lower metal temperatures and significant savings in cost. Gas turbines can also be optimised around stage efficiency and capital cost with any resultant increase in exhaust temperature T_2 being taken as useful energy into the steam turbine.

2.6 Combined heat and power

Whilst there have been continued improvements in the technology and efficiency of power generation over the last few years and despite the significant step change improvement in efficiency achieved by combined cycle plant, the energy from over 40% of the fuel consumed is still wasted, rejected as low grade heat to the atmosphere. In addition it has generally involved consumption of a prime fuel (natural gas) which was until recently regarded as too valuable a commodity to use in power generation.

Power supply and demand forecasts within the National Grid Company's "Seven Year Statement" imply a peak of approximately 37,000MW of generation from fossil fuel fired power plant during the year 2000. Based on the forecast mix of generating plant at that time, this implies up to 42,000MW of heat derived from fossil fuels being rejected to the atmosphere and lost.

In parallel with this large and thermodynamically inevitable rejection of waste heat associated with pure power generation, prime fossil fuels are being used to generate heat directly for industrial process use and for space heating of commercial and domestic buildings. Used for heating alone, the work-producing capability of these fuels, the potential to generate power, is entirely wasted.

The thermodynamic and environmental case for CHP is that, in generating heat (be it for industrial, commercial or domestic use) from precious fossil fuels, the opportunity should always be taken to generate power in the process.

2.7 The science of steam and gas turbine CHP cycles

The key factor in the selection and configuration of components for a CHP plant is the ratio of heat and power required by the end user.

The so-called Total or CHP efficiency is simply defined as:

$$\eta_{CHP} = \frac{\text{Power output} + \text{useful heat output}}{\text{Fuel input}}$$

Whilst this is clearly a good indication of energy efficiency, it takes no account of the higher value, thermodynamically and economically, of electrical power compared to heat. Without going further into the thermodynamics and the concept of work-producing capability (exergy), it is sufficient to observe that electricity generally costs at least four times as much as heating fuel per unit of energy. It is evident that η_{CHP} alone is not a good indicator of the merit of a CHP plant, the relative proportions of power and heat produced are crucial. The trend has been towards CHP plants which maximise the power generated and minimise the heat to power ratio, as outlined below.

Steam turbine CHP plant

Steam turbine CHP (fig. 5) may be configured with back pressure, extraction condensing or pure condensing turbines according to the ultimate form in which the useful heat is to be delivered. For industries requiring process or heating steam this is a back pressure type. For applications such as district heating requiring medium temperature hot water, it could be a condensing type. For a typical industrial application the pressures at which steam is required for process use, combined with practical limitations on boiler steam temperature and pressure, mean that the generated heat to power ratio is about 5 or greater. The η_{CHP} may be high (around 90% is achievable) but the opportunity to generate power is limited.

Gas turbine CHP plant

A simple gas turbine CHP plant, where the turbine exhaust gases are passed via a waste heat recovery boiler to generate steam (or hot water, or other heated process fluid) provides a significantly reduced heat to power ratio, between about 1.3 and 2.5. The lower end relates to turbines, normally aeroderivative units, with high power generation efficiencies.

Combined cycle CHP plant

Figure 6 shows the combined cycle gas turbine system integrated into a CHP plant. The example chosen represents a common industrial CHP configuration, with a back pressure steam turbine. For gas turbine based CHP, the inclusion of a steam turbine maximises the power generated for a given quantity of useful heat required. Using an efficient gas turbine with a back-pressure steam turbine the heat to power ratio can be reduced towards 0.7 and η_{CHP} remains above 80%.

Thus the combined cycle gas turbine CHP plant can provide a substantial proportion of its output as power, whilst retaining high energy efficiency.

2.8 The practice of gas turbine CHP engineering

The selection and configuration of equipment for a CHP plant requires a broader view than pure power station design. In general terms it is advisable to treat the CHP plant as a process plant from which electrical power is a desirable by-product.

For a natural gas fuelled industrial gas turbine CHP plant producing process steam (substantially the most significant CHP plant type in terms of installed capacity) the key design issues are:

Provision of steam at the required conditions, normally over a wide demand profile

Matching plant operation to a varying heat to power ratio
Dealing with dynamics in process steam demand
Maintaining high security of steam supply
Utilising returned condensate from the process
Integration with existing steam systems.

The above must be controlled to maximise operating efficiency and dependability in accordance with an agreed "control philosophy", this normally demands full automation.

To be operationally successful, the CHP plant must first and foremost meet the above requirements. While the gas turbine(s) are selected from standard products, the waste heat recovery boiler and steam cycle design will be bespoke and potentially complex, with provision for heat bypassing or dumping, supplementary boiler firing and fuel diversity.

The range of configurations is evident in CHP plants built by Balfour Beatty, two of which are illustrated in figures 7 and 8. The first of these, based around twin 4.5MW gas turbines, employs two waste heat recovery boilers with capability for natural gas supplementary firing, supported by a conventional boiler with dual fuel capability. The boiler arrangement caters for the full range of heat load, plant outage and fuel availability conditions. The plant is unmanned and automated to meet a complex control philosophy. The second plant is based around one 40MW gas turbine and waste heat recovery boiler, operating in combined cycle with a steam turbine. In this case the major complexity is associated with the steam turbine which, to maximise power generation, has a total of four admission or extraction points plus a condensing section, and is backed up by multiple steam bypasses for security of process steam supply.

3. THE DEVELOPMENT OF CHP

3.1 Technology
In the UK, CHP historically has been used principally in energy intensive industries such as petrochemical, paper and board, foodstuffs and iron and steel.

Limited availability of alternative technology and the lack of availability of economic gas or light oil fuels dictated that the great majority of plant installed before 1985 comprised coal or heavy oil fired steam boilers and back pressure or extraction condensing steam turbines.

Historical survey data on CHP is limited, but CHP capacity declined in the UK from the 1960's through to the late 1980's. This capacity reduction principally involved decommissioning of steam turbine CHP plant which can be attributed to many factors, including a decline in heavy industry, security of supply from the national grid, shifts in the relative prices of fuel and electricity, improvements in industrial boiler technology and pressures on manning levels.

More recent survey data collated by ETSU (figure 9) shows the subsequent growth in CHP. This capacity growth principally involves gas turbine based industrial CHP plant using the now readily available and economic natural gas as fuel, which as discussed above, has a better match to the lower heat to power ratios that normally apply today. Figure 10 shows the distribution of types of plant used by CHP schemes in the UK.

3.2 Government Policy
The profile and uptake of CHP for industrial and non-industrial use before the 1980's was limited in the UK and elsewhere. Government policy and the structure of the often nationalised and invariably heavily regulated electricity supply utilities was geared towards

the security of supply and efficient electricity generation which generally discriminated against CHP. Notable exceptions were in certain former Eastern Bloc countries where centralised CHP schemes (usually coal fired) providing district heating were supported and in the Netherlands where gas fired CHP has been promoted by government energy policy since the early 1970's.

The factors affecting growth of CHP can broadly be grouped into four categories [1]:

> Electricity and fuel industry structures and attitudes
> Government policy
> Electricity and fuel prices
> Consumer demand profiles

Within the UK, a marked change in the first two of these categories was signified with the privatisation of the electricity and gas supply utilities during the late 1980's and early 1990's. In particular this released restrictions on natural gas as a prime fuel, which could now be consumed in bulk for power generation and purchased competitively from a range of suppliers.

In parallel with this, UK Government policy relating to the environment recognised the potential of gas fired power generation and CHP. Most well known is the commitment made in the 1990 White Paper "This Common Inheritance", to return growing UK Carbon Dioxide emissions to the 1990 level by the year 2000. Whilst the UK government has not adopted all available measures to reduce CO_2 emissions (in particular it continues to exercise a veto on Carbon Tax) it has set targets for the installation of CHP, the current target being 5000MWe installed capacity by the year 2000.

In the late 1980's and early 1990's falling gas prices and rising electricity prices favoured CHP. Recent volatility in both gas and electricity markets has made development more difficult, in particular "spot" gas prices varied by a factor of two during 1995. As part of government policy CHP was excluded from the fossil fuel levy to encourage CHP development, but the fossil fuel levy has now been substantially reduced, from 11% to 2.2% .

Government commissioned reports conclude that "CHP remains the most important single technology as far as potential energy saving measures are concerned".

3.3 Economics of gas turbine CHP

The cost of building gas turbine CHP plant, whilst less than coal fired or nuclear power generation, can be significant. Depending on the scale, complexity and heat output, costs for new plant range from £0.6 million to £1 million per megawatt of electrical output, and this does not compare favourably with CCGT plant at under £0.5 million. Fortunately, the savings in operating cost are significant too, as the effective cost of power generated on site via CHP is low compared to the avoided cost of imported power. Internal rates of return around 15% are commonly achieved over a typical 15 or 20 year write-down period.

Notwithstanding the adequate investment returns which can be achieved, CHP projects have to compete for funds with other capital projects. Often projects related to the industrial process itself may show shorter term returns, or lower commercial risk, or simply be judged more relevant to the "core business". Thus whilst the environmental benefits of CHP plants are immediately apparent the commercial benefits may be more difficult to realise. This is especially true when the market place for both gas and electricity is unstable.

These circumstances have seen the new build of gas turbine CHP plant dominated by Contract Energy Management (CEM) agreements. The scope of CEM is flexible but typically, the CEM company will build, own, operate and maintain the plant, providing heat and electricity as a complete energy package against a fixed term tariff scale.

3.4 Quality of Life
The theme of this symposium is "Improving the quality of life through technology". CHP achieves this, directly and indirectly in a number of ways:

Emissions
The combustion of fossil fuels for power and heat generation gives rise to emissions of carbon dioxide (CO_2). Other gaseous pollutants may also be generated, depending on the composition of the fuel and the combustion process.

Increasing CO_2 levels in the atmosphere contribute to the "Greenhouse effect", associated with global climate change. CO_2 is, however, produced in large quantities by combustion, since the fundamental aim is to convert all the available carbon to CO_2 to release its chemical energy. The energy efficiency of CHP plant is, therefore, very important in any strategy to control CO_2.

Oxides of sulphur (SO_x) in flue gases produce acidic compounds in the atmosphere leading to "acid rain". SOx is produced by the combustion of sulphur in fuel in proportion to the fuel's sulphur content. Natural gas has negligible sulphur content, virtually eliminating SO_x. It can, however, also be reduced after combustion by flue gas de-sulphurisation, but the economics of this are such that its application is generally limited in the UK to the largest power stations and the combustion of specific fuels e.g. waste incineration.

Oxides of nitrogen (NO_x) damage the ozone layer thus contributing to the greenhouse effect and also cause "acid rain". NO_x can also lead to "smog" formation under certain atmospheric conditions. NO_x is produced from the combustion of fuel-bound and atmospheric nitrogen. Unlike CO_2 and SO_x, the formation of NO_x can be limited by control of the combustion process.

Figure 11 shows the CO_2 and NO_x emissions when power and heat are provided to a typical industrial process by a gas turbine CHP plant compared with other typical forms of separate power and heat production. The figure is based on statistical and sample data from the Energy Efficiency Office[2].

Power generated by CHP will tend to displace generation from older low merit power stations. In the typical case, CO_2 emissions would be reduced by over 60% and NO_x emissions by over 70%.

Fuel conservation
As explained earlier in this paper, CHP represents the most efficient way to utilise fossil fuels for heat and power generation. Even when compared to state of the art CCGT power generation plant and modern industrial boiler or process heating plant, CHP can reduce fuel consumption by around 15%.

Land use
The sites chosen for CHP in industry tend to be "brownfield" sites, either the site of a former process or utility plant or an available area within a larger complex. All recent Balfour Beatty projects have been built upon such sites.

Reliability and Security

When CHP plants are planned, secure electrical supply is provided by continued, parallel, connection to the grid network. As a result of providing two diverse sources, security of supply can be dramatically increased. This can be a significant benefit where power interruptions are an issue, for example in hospitals or continuous processes. Outside the UK, in developed and developing countries where electricity demand outgrows supply, the installation of CHP by industry and the public sector can help balance the transmission network and reduce the incidence of blackouts and "brownouts".

Economic benefits

CHP provides an economic benefit; minimising energy charges, which on average account for around 13% of industrial operating costs.

4. PREDICTIONS FOR THE FUTURE

Improving technology, increasing environmental constraints, and continual pressure to minimise costs within industry and across the marketplace, imply a strong potential for gas turbine CHP and for CHP in general. Energy efficient distributed power generation, with gas turbine CHP plants serving communities and business areas, is a perfectly feasible prospect given suitable market conditions and policy.

Taking a broader view than conventional gas turbine or engine based CHP there is also a wide range of technologies with potential for CHP development, including:

Waste fired CHP plants - these are established in Europe, particularly burning municipal waste and providing district heating, or burning process waste such as wood bark and providing process steam, but have little presence in the UK market, even though waste to power plants have been promoted by the Non-Fossil Fuels Obligation.

Fuel cell CHP - at present fuel cells operate on natural gas fuel at over 40% power generation ("conversion") efficiency with potential for heat recovery. Higher temperature cells are being developed with conversion efficiencies of 55% and high grade heat recovery, with module sizes in excess of 10MW.

Micro-CHP - conceptual designs exist based on very small scale gas engine generators integrated with heating systems in the home, providing domestic power in parallel with the grid supply when the heating boiler is operating.

The UK government recognises the environmental benefits of CHP and it is hoped that positive action will be taken to reach declared targets for installed CHP capacity. As noted earlier, however, the current volatility of fuel and electricity prices is a deterrent to the capital investment required. Other deterrents, particularly for non-industrial CHP, include the high costs of infrastructure for heat distribution. If the market place is left to its own devices, it is unlikely that any of the above ideas, or indeed the fullest implementation of CHP, will occur.

If the benefits to the environment and energy conservation offered by CHP are to be realised, clear public policy initiatives will have to be produced. These could take the form of: an environmental levy - to reward energy efficient generation and reduced environmental pollution, differential pricing - to provide a stable investment case for CHP, and planning regulations - to ensure that in all developments the installation of CHP, or appropriate infrastructure such as heat distribution systems to enable CHP, is a necessary condition.

References

[1]   "The impact of liberalisation of the European electricity market on cogeneration, energy efficiency and the environment". Ramboll, Denmark & Ilex Assoc., UK, COGEN Europe/E.C. Study, March 1997.

[2]   "Environmental aspects of large scale combined heat and power". Energy Efficiency Office GPG116, August 1994.

[3]   "Industrial sector Carbon emissions database and model for the UK" ETSU, 1997.

The authors would also like to acknowledge the assistance provided by the CHPA and colleagues at Balfour Beatty in the preparation of this paper.

FIGURE 1 - THE HEAT ENGINE

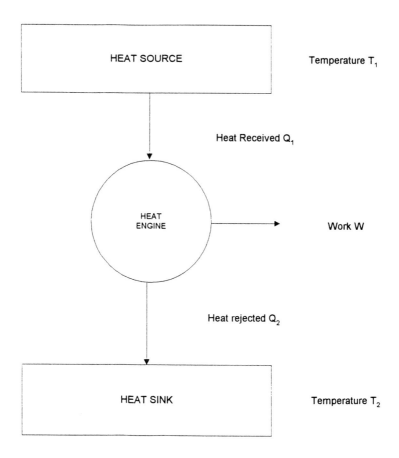

CONDENSING STEAM TURBINE POWER PLANT
FIGURE 2

a) SCHEMATIC

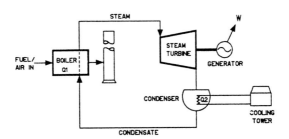

b) CYCLE

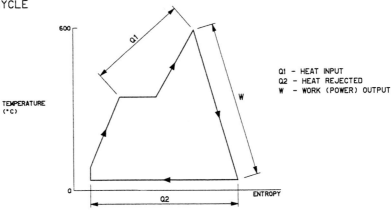

Q1 – HEAT INPUT
Q2 – HEAT REJECTED
W – WORK (POWER) OUTPUT

c) TYPICAL HEAT BALANCE

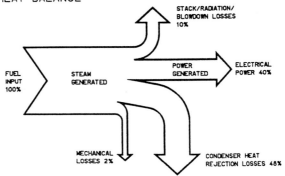

TOTAL USEFUL OUTPUT : 40%

OPEN CYCLE GAS TURBINE PLANT
FIGURE 3

a) SCHEMATIC

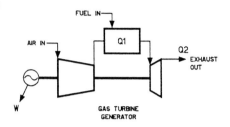

b) CYCLE

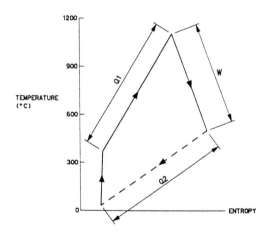

c) TYPICAL HEAT BALANCE

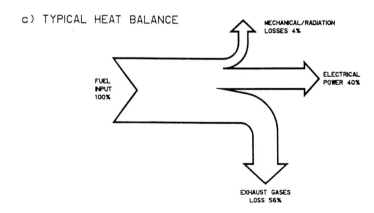

TOTAL USEFUL OUTPUT : 40%

COMBINED CYCLE GAS TURBINE PLANT
FIGURE 4

a) SCHEMATIC

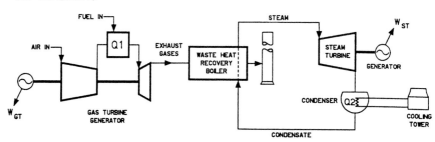

b) CYCLE

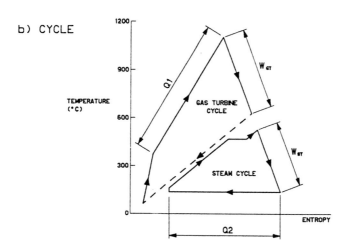

c) TYPICAL HEAT BALANCE

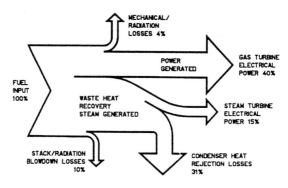

TOTAL USEFUL OUTPUT : 55%

BOILER/STEAM TURBINE CHP
FIGURE 5

a) BACK PRESSURE TURBINE

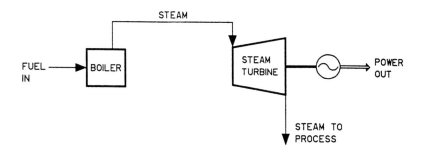

b) EXTRACTION CONDENSING TURBINE

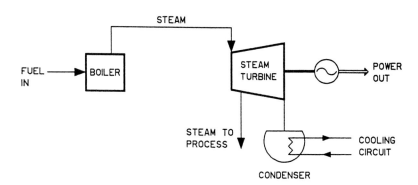

COMBINED HEAT AND POWER PLANT
FIGURE 6

(GAS TURBINE/WASTE HEAT RECOVERY BOILER/
BACK PRESSURE STEAM TURBINE CONFIGURATION)

a) SCHEMATIC

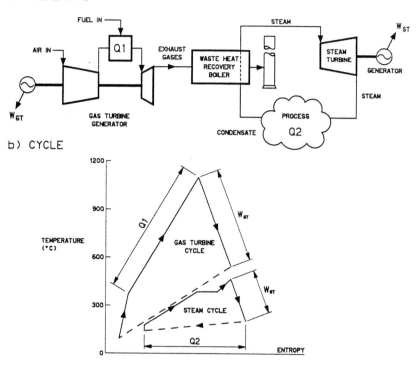

b) CYCLE

c) TYPICAL HEAT BALANCE

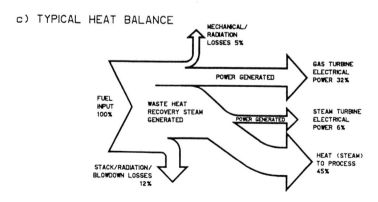

TOTAL USEFUL OUTPUT : 83%

Fig 7   CHP scheme built by Balfour Beatty based upon 4.5MWe gas
turbines

Fig 8   CHP scheme built by Balfour Beatty based on 40MWe gas turbine
(similar to plant described in text)

FIGURE 9 - TRENDS IN INSTALLED CHP CAPACITY

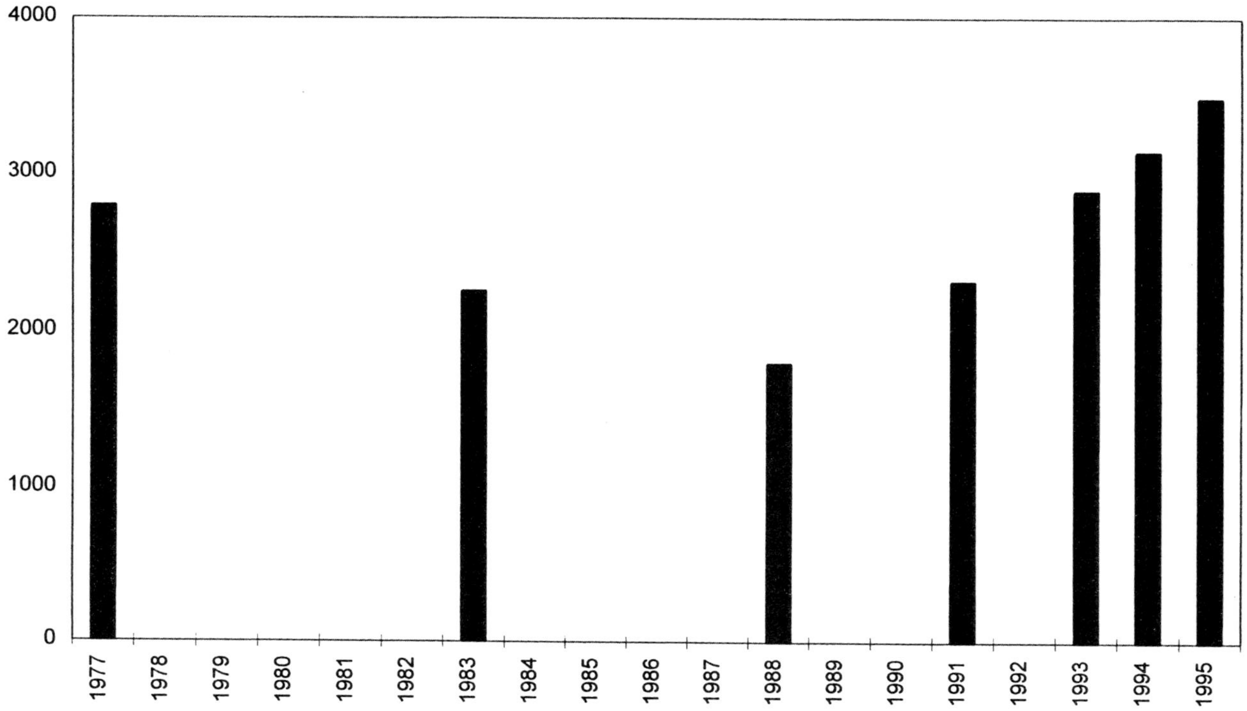

SOURCE - ETSU DATA

128

C527

FIGURE 10 - TYPES OF PLANT AND FUELS IN UK CHP SCHEMES

a) Type of plant

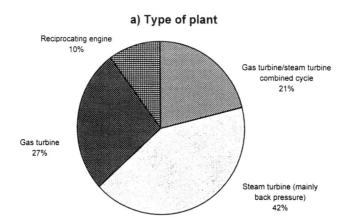

b) Types of fuel

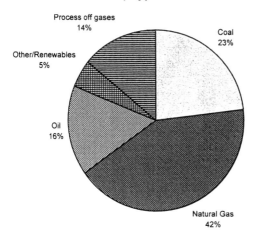

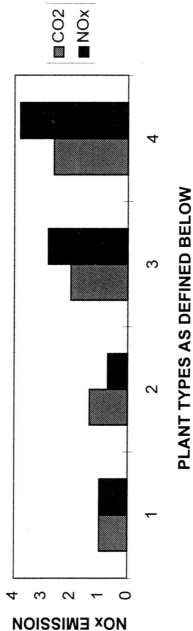

FIGURE 11 - RELATIVE EMISSIONS

CO2
NOx

RELATIVE CO2 or NOx EMISSION

PLANT TYPES AS DEFINED BELOW

1 - Gas turbine CHP plant supplying power and heat
2 - Power via grid from CCGT power station, heat from on-site gas fired boiler plant
3 - Power via from grid (UK mix of power plant), heat from on-site oil fired boiler plant
4 - Power via grid from coal fired power station, heat from on-site oil fired boiler plant

sign considerations for a hybrid heat/engine electric
hicle

essor **J N RANDLE** FIMechE FEng, RDI, FRSA
tor, Volvo Aero Turbines (UK) Limited and Director, Automotive Engineering Centre, School of Manufacturing
Mechanical Engineering, University of Birmingham, UK

The paper considers the options that are available to improve the emissions characteristics of

motor vehicles within known or predictable technologies. It discusses how, the way in which

vehicles are operated affects the design solution, what emissions improvements over present

technology could be expected and how that could influence total vehicle emissions by the year

2010. The paper details two specific designs, one for a taxi the other a privately owned, small

general purpose vehicle.

SCIENCE AND THE MOTOR CAR

More than any other piece of mechanical engineering, the motor car has impacted the lives of
everyone on the planet. It has brought about enormous improvements in the quality of life and
has provided millions of people with a livelihood, but it has been at a cost; a cost in lives lost
through accidents, loss of precious countryside buried beneath thousands of miles of tarmac,
pollution of the air we breathe and now a threat to the global environment. The twentieth
century has been the age of the car, the twenty first century must be one in which the negative
impacts of the motor car are as far as possible eliminated and where a proper balance between

public and private transport is achieved. The coming century will bring new discoveries, inventions that may very well provide complete answers to the problems that now beset us, but as engineers and scientists we can only work within the envelope of that which we know and that which we can reasonably predict.

AUTOMOBILE ENGINEERING INTO THE TWENTY FIRST CENTURY.

Now that the argument over whether or not global warming exists has been settled and there is agreement that it does exist and that its effects could be such as to raise the average temperature of the Earth by between 2 and 3 degrees Celsius over the next 100 years, the priority for vehicle designers is clear; we must reduce the emission of carbon dioxide. There are many gases which influence the greenhouse effect but technology exists to bring them under control, whereas carbon dioxide, being the stable, fully oxidised consequence of the combustion of a hydrocarbon fuel, only reverts to carbon and oxygen through biological processes and cannot be converted through exhaust after-treatment. It follows therefore, that the automobile engineer's task is to reduce production by reducing fuel consumption or by reverting to hydrogen rich fuels. The specific fuel consumption of engines has improved very little over the last thirty years; paradoxically, improvements have been to a large degree inhibited by emissions regulations, which require catalytic after-treatment, which in turn demands a constant stoichemetric air/fuel ratio. On the assumption that this is the ongoing technology, the size of the task is daunting.

TARGETS

In a paper delivered at the 1993 Festival of Science, the writer presented data on what was required to meet the targets set by the 1992 Rio world summit. To play its full part, the motor industry needed to reduce the average fuel consumption of all vehicles in the market place by 23% by the year 2000 and to achieve that those vehicles needed to be on sale at that time.

The average fuel consumption for all the models on sale at that time was 25.6 m.p.g., the worst being 12.1 m.p.g. and the best 54.7 m.p.g.

For 1996 the figures were 25.9 m.p.g., the best being 47 m.p.g. and the worst 11 m.p.g. At best a 1% average improvement over 4 years! Since it takes 10 years to replace all of the cars in the market place, it is reasonable to assume that the Rio target will not be met by the vehicle fleet.

The same paper extrapolated the requirement to 2010, assuming only the requirement to maintain 1990 emissions levels; that demanded a CO_2 reduction of 38%. Now that the requirement has been stiffened to achieve a 10% reduction against 1990 levels, i.e. a reduction of 18 megatons, which if equitably distributed, would require the vehicle fleet to accept responsibility for 3.8 megatons, the average reduction across the whole of the transport industry becomes 48%. A daunting task with today's technology, when measured against recent performance.

POSSIBLE SOLUTIONS.

There is no simple all embracing answer to the problem, but there are solutions that are most appropriate to the vehicle that is being designed. The types of vehicles and the way in which they are to be used (referred to hereafter as the duty cycle), determines the design solution. The duty cycles that predominate are:-

1. Motorway and rural driving.

2. Routes into and out of cities and towns along radials to the centre.

3. Circulatory routes within the city.

Privately owned passenger cars largely fit into groups 1 and 2. Buses and collection and delivery vehicles are mostly group 2, whilst taxis occupy group 3. The principal characteristics of the duty cycle that influence the design solution are the way in which the engine is used and the way in which the tractive energy is used, i.e. for acceleration, for hill climbing, to overcome aerodynamic drag, to overcome tyre losses or to overcome frictional losses:-

Motorway and rural driving.

Since this duty cycle is predominantly at constant high speed, it has the following characteristics *relative to the overall energy requirements:-*

- high aerodynamic force.
- high tyre losses.
- low frictional losses.
- low accelerative force.
- no idling losses.
- engine and transmission operating near maximum efficiency.
- not operating in emissions sensitive area.

City driving radial routes

Radial city routes follow a mixture of city and urban driving giving *relatively:-*
- low aerodynamic losses.
- low tyre losses.
- high accelerative losses.
- high frictional losses.
- medium idling losses.
- engine operating at medium efficiency.
- operating regularly in emissions sensitive area.

City driving circulatory routes

Circulatory city routes involve continuous heavily trafficked conditions giving *relatively*:-
- low aerodynamic losses.
- low tyre losses.
- high accelerative losses.
- high frictional losses.
- high idling losses.
- engine operating at low efficiency.

operating continuously in an emissions sensitive area.

TECHNICAL OPTIONS.

There are relatively few options available to the designer within those technologies that are relatively mature, or which are likely to become mature, within say the next thirty years. They can be described as follows:-

Conventional transmissions in lightweight body structures.

Considering the vast investment in today's technology, reducing the weight of the vehicle is, with the exception of using other fuels, the most logical and least disruptive route for the industry to follow in the short term. Reduced weight improves accelerative and tyre losses but does not effect aerodynamic losses unless the size of the vehicle is reduced as a consequence. Furthermore, it does not achieve a pro rata improvement in frictional losses, therefore its effect upon steady high speed consumption is less marked than upon the slow speed cycles involving constant acceleration and deceleration. Also since conventional drive-trains give the best high speed economy of all of the near term options, there is strong pressure to follow this route. However, because the effects relative to the duty cycles imposed are variable, historical rule of thumb experience predicts that overall fuel economy will only improve to around 50% of the reduction in weight, i.e. a 50 % weight reduction in an otherwise unchanged private passenger vehicle would bring about a 25% economy improvement. A recent survey of the overall fuel consumption of similar vehicles of varying weights, see Figure. 1, gives some support to this rule but underlines the importance of weight on the urban cycle where for the reasons outlined above, the relationship is much higher at around 90%. The problem is twofold, legislation is driving towards higher safety standards which demand such features as side intrusion members and the market place has become educated to expect more and more sophistication. The result has been that cars have become heavier and that steadily increasing trend shows little sign of reversing. The following table is a random selection of cars in the market place in 1989 and their successors in 1996, showing kerb-weight increases.

'89 BMW 530	1510 kg	'96 BMW 528	1530 Kg
'89 Ford Fiesta. 1.6	890 Kg	'96 Ford Fiesta. 1.4	978 Kg
'89 Citroen BX. 2L	1000 Kg	'96 Citroen Xantia. 2L	1299 Kg
'89 Peugeot 405. 1.9	1020 Kg	'96 Peugeot 406. 2L	1300 Kg
'89 Opel Vectra. 2L	1255 Kg	'96 Vauxhall Vectra. 2L	1300 Kg

With the wide range of options available to the customer, it is difficult to be precise about the rate of vehicle weight increase, but it is probably in the order of 1% per year across the whole of the vehicle fleet.

Weight v Fuel Consumption

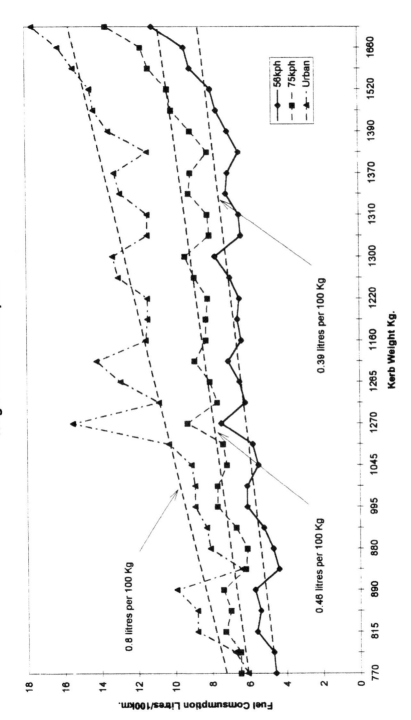

Figure 1.

Kinetic Energy or Electrical Energy storage systems.

None of the systems that exist, or seem at this time to be possible, have anything like the storage capacity of liquid fuel; Figure. 2 summarises the position. Even the best available mature technology battery is around 100 to 1 inferior to petrol and it is doubtful that any battery technology will better 30 to 1. Kinetic energy storage is presently no better, at around 120 to 1.

It is a popularly held belief that an electric vehicle uses more fuel and creates more emissions than its conventional cousin, but does so at a location remote from emission sensitive areas. Although it is true that an electric vehicle, if used in a rural and motorway duty cycle, runs at a lower overall efficiency than a conventional vehicle, this is not the case if it is operated in a city environment where its ability to completely close down when the vehicle is stationary and to run in a high efficiency mode when the vehicle is trickling along, is actually quite good, as the paper noted earlier concluded, see Figure. 5. It follows therefore, that battery electric vehicles will only be viable under very specific, largely city based, low range, low performance, duty cycles and will probably never be able to compete with conventional vehicles in general use.

Motor technology.

A number of technologies are available to provide traction for the hybrid or pure electric vehicle, namely:-

- Wound field, brush commutated DC.

- Slip ring, AC induction (asynchronous).

- Brushless DC (synchronous).

- Switched reluctance (synchronous).

Motor speeds are low relative to the generator, therefore a *brushed commutator DC motor* cannot be ruled out. It has after all been the mainstay of electric traction vehicles since their inception. It does however, suffer a number of problems which mitigate against its use. Commutation losses are high, it is heavy relative to the other types, it creates a great deal of radio frequency interference and the carbon dust that it generates causes arcing and thereby reduces the life of the motor.

The *induction motor* is a well tried and robust technology. However, its use with high frequency variable speed drives, does require careful attention to the design and selection of materials. It is able to operate at high temperatures and with the use of vector control, can provide the required envelope of operation. Its overall efficiency is good and it is significantly lighter than a brushed DC motor. Efficiency falls off at high speeds and loads. It has a reasonable torque density and is relatively cheap to produce.

The *brushless DC motor* is a confusing name; the machine is in fact an AC synchronous, permanent magnet motor with rotor position feedback, such that the characteristics when looking at the DC bus, represent the characteristics of a DC shunt motor. This provides a motor with a field system which is a mixture of permanent magnets and armature reaction from vector control. This machine is very efficient in the cruise mode, has an adjustable field, is robust and relatively easy to make.

The *switched reluctance motor* is a relative newcomer. It operates in a similar manner to the switched DC permanent magnet motor, but has lower specific power. It has the potential of reduced cost over the switched DC motor but with higher control costs.

The relative characteristics of each motor are shown in Figure 3. At the present state of development, the switched DC motor is the most attractive and was chosen for the two vehicles described later.

AVAILABLE BATTERY TECHNOLOGY CHARACTERISTICS

TYPE	ENERGY DENSITY Wh/Kg	POWER DENSITY W/Kg	COST £/kWh	CHARGE FACTOR Ch in/Ch out	LIFE Deep cycles	CHARGE TIME Hours	OPERATING TEMP Degrees C
Lead/acid	24-40	60-100	260	1.05	1000	8	+5 to +50
Adv. Lead/acid	50	150	?	1.05	1000	<1	Ambient
Ni/Cd	40-50	175	870	1.2	2000	6.5	-20 to +45
Na/S	80-120	90-105	700	1	>1500	10	+290 to +340
Na/NiCl$_2$	100	110	1100	1	1000	10?	300
Zn/Br	65	70-100	870	1.15	>1500	7	Ambient
Metal Ni Hydride	55	100	800	?	>500	11	Ambient
Lithium polymer	200	150	800	?	?	10	Ambient
Lithium iron sulphide	100	150	?	?	500	10	450

Figure 2.

AVAILABLE MOTOR TECHNOLOGY

Characteristic	DC	AC induction	Switched DC	Switched reluctance
Torque/weight	low	low	high	medium/high
Power/weight	low	low	high	medium/high
Efficiency	medium	low	high	probably high
Comm losses	high	none	none	none
Dyn behaviour	bad	good	good	good
Life	medium	high	high	high
Cooling difficulty	high	high	low	low
Shape flexibility	low	medium	high	medium
Development cost	low	low	high	high
Motor cost	high	low	high	low
Control dev cost	low	high	medium/high	high
Control cost	low	high	medium/high	high

Figure. 3

FUEL CELLS

The fuel cell has the potential for solving the problem; in simple terms it combines hydrogen with oxygen to form water, releasing electrical energy in the process. It is highly efficient; 50 to 80% dependent upon the type of cell used and is already seeing successful application in stationary generating equipment in the USA where the majority of the research into this technology is being conducted. For vehicular use however the technology presently has serious drawbacks, namely:-

Weight and volume-Although the fuel cell itself may have tolerable energy density, the ancillary equipment required to pump air and to convert the on-board fuel to hydrogen, if this is the system chosen, is very heavy and voluminous, see Figure. 4.

Operational characteristics-Start-up times are slow, as are demand responses, which in turn demand on-board electrical storage to provide transient power, creating a further substantial weight penalty.

Provision of Hydrogen-To be truly pollution free, fuel cells require hydrogen provided from a renewable source and an infrastructure to deliver it, neither of which is likely to be available within the time scales which would be demanded by the design and development of a vehicle which could be started now and which would influence the environment within the time scales under discussion. Present designs suggest the use of methanol as the hydrogen bearing fuel which would be reformed on board, but which would release CO_2 in the process. This would be acceptable in terms of local pollution control, but would still impact the global warming problem.

FUEL CELL TECHNOLOGIES (Neglecting Steam Reformation).

	PEMFC	AFC	PAFC
Operating Temp (C)	80	70	190
Air Pressure (Bar)	3.5	0.04	3-8
Air Requirement	3 x Stoichemetric	2.5-3 x Stoichemetric	3 x Stoichemetric
System Volume	3.7P+F(170+19P)	17P+F(50+5.7P)	7.5P+(50+5P)
System Weight	5P+119 v'P	15P+80 v'P	11P+64v'P
Power/Sq.cm	500 mA	300 mA	200 mA
DATA FOR TYPICAL 100 kW UNIT.	2440 Litres 1690 Kg	2330 Litres 2300 Kg	1200 Litres 1740 Kg

PEMFC   Proton exchange membrane fuel cell.   P= Power kW.
AFC   Alkaline fuel cell.   F= Packaging factor normally 2 to 3.
PAFC   Phosphoric acid fuel cell.

Figure. 4

ALTERNATIVE FUELS.

The paper noted earlier evaluated the options available, including methanol, ethanol, hydrogen and methane, see Figure. 5, and concluded that methane in the form of natural gas was the most logical, freely available alternative fuel. The key advantage of methane is its high hydrogen content, which means that for a given energy release it produces around 25% less CO_2 than other hydrocarbon fuels and is available in very large quantities. It does however have the following disadvantages:-

• Unburned methane present in the exhaust gases is 20 to 30 times worse than CO_2 as a global warming agent. This can very easily overwhelm the advantages noted above.

• Converting methane by exhaust after-treatment is difficult, since the combination of cooler exhaust temperatures (around -50 deg Celsius) and the higher catalyst operating temperatures required (around +300 deg Celsius), present difficulties in start-up and conversion efficiency that have not yet been resolved at a commercially available level; although it is expected that this will be achieved within the next year or two.

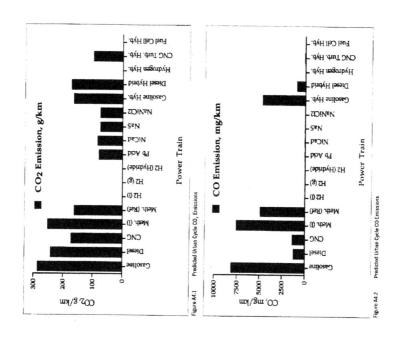

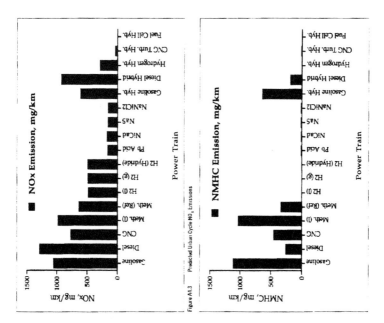

Figure 5.

NOVEL ENGINES.

Intermittent combustion engines.

The principal problem associated with conventional internal combustion, reciprocating engines is that through being forced by legislation down the catalytic after-treatment route, they are confined to the air/fuel ratios that catalysts require to operate effectively, which in turn puts a ceiling on the improvements in economy and therefore reductions in the production of CO_2 that can be expected, so that improvements under motorway and rural conditions have to be achieved by attention to the vehicle itself. A way around this problem is the dual operation engine which reverts to a lean burn combustion cycle when operating at higher speeds. A good example of such an engine is Mitsubishi's direct injection gasoline engine, which is claimed to give similar fuel economies to a small diesel. This, together with closer control of the combustion process within individual cylinders, probably describes the best that is likely to be achieved from these engines and may represent a 10% overall improvement over conventional engines. Direct injection small diesels using variable injection phasing, will probably deliver a similar improvement. Catalyst warm-up via electrical or fuel injection will increase fuel consumption marginally.

Catalytic combustion chambers, which would overcome the start-up problem, do not seem to be possible.

Continuous combustion engines.

Continuous combustion engines have significant advantages in terms of emissions, for the following reasons:-

- There is no freezing of the chemical activity, hence complete combustion can be assured.
- Combustion temperatures can be maintained within close limits, hence NO_x production can be limited.
- Starting enrichment can be limited.
- Catalytic combustion chambers are theoretically possible, ensuring fast start-up.
- They are very suitable for natural gas, effectively eliminating the problems outlined earlier.

Two engines can be considered; the Stirling engine, featuring an external combustion chamber providing heat to a closed cycle reciprocating engine and the gas turbine. The Stirling engine tends to have a relatively poor power to weight ratio because of its low speed of operation and despite many attempts, has not yet seen successful application in an automobile. The gas turbine has also failed to reach production, but with the introduction of hybrid drive-trains the reasons for its lack of success may now be close to resolution.

The case for the gas turbine.

Many researchers have put the case for the gas turbine, (R.W. Chevis, J Everton) Autotech '93. (M.R.Jones, T.Sawyer) I. Mech. E paper C462/45/138 Autotech '93. (K.R.Pullen) and (J.N.Randle and D.E.Embrey) Battery Electric and Hybrid Vehicles. I. Mech.E Seminar '92.

Gas turbines have been around for a very long time; they are very well understood, have legendary reliability and yet they have not found favour for automobile use. The reasons for this are:-

• Unlike a reciprocating engine that can develop a range of torque for a given speed, gas turbines are principally a speed related device, hence power is not instantaneously on tap - there is a lag. Furthermore, their power delivery is almost identical to the vehicle power requirements, see Figure 8, so, when in the gear that is required for maximum speed, there is no excess of power for acceleration purposes unless a lower gear is selected. A gas turbine needs a large number of intermediate gear ratios or a wide ratio Constantly Variable Transmission (CVT).

• There is little overrun torque, hence driveability is less precise.

A small gas turbine driving a small, high speed, direct drive generator providing power for batteries and a traction motor, is however much more attractive. By using the gas turbine as a high speed generator, which for a significant part of its operating cycle is at constant speed and therefore at high power relative to the road load requirements, most of its shortcomings are overcome. Even in the parts of its cycle where it is running at relatively low power; flowing energy into and out of the batteries to provide extra power, thus eliminating lag and any other transient unattractive characteristic, is perfectly practical. It is even possible to use the generator as a motor to accelerate the gas turbine to a higher speed, thus eliminating any extra emissions brought about by rich running during acceleration. Electric traction motors deliver constant torque over the lower part of their operating rev range, then constant power up to their speed limit. They are therefore more attractive than mechanical transmission internal combustion engines of any type, see Figure 8. The gas turbine has the following advantages/disadvantages relative to conventional drive-trains:-

Advantages:

• Being a constant combustion device, there is very precise control over the combustion cycle. Emissions are much better than a gasoline or diesel engine, even without the use of catalytic after-treatment, see Figure 9.
• Gas turbines do not have the cold start-up problems that catalyst equipped gasoline engines have, nor the start-up smoke problems of diesels.
• They have a soft start - very important when operating to a hybrid cycle.
• Small gas turbines using recuperator heat exchangers are very efficient, up to 34% thermal efficiency, against 32% for a small diesel and 26% for a gasoline engine.
• The very high operating speed of a gas turbine when driving a direct drive alternator, ensures a very light generator set.
• It is a true multi-fuel engine.
• It requires no external cooling system.
• It has very low maintenance requirements.
• It is very quiet in operation.

Its disadvantages are:

- It is presently very expensive.
Heat exchangers, necessary to raise the thermal efficiency, require significant development.

GENERAL ARRANGEMENT OF A SINGLE SHAFT GAS TURBINE.

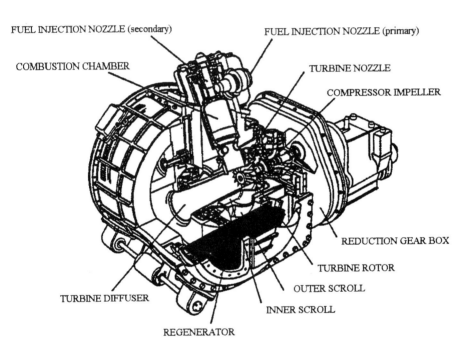

FUEL INJECTION NOZZLE (secondary)   FUEL INJECTION NOZZLE (primary)

COMBUSTION CHAMBER   TURBINE NOZZLE

COMPRESSOR IMPELLER

REDUCTION GEAR BOX

TURBINE ROTOR

OUTER SCROLL

TURBINE DIFFUSER

INNER SCROLL

REGENERATOR

Figure 6.

EFFECIENCY CHARACTERISTICS OF GAS TURBINE TECHNOLOGIES

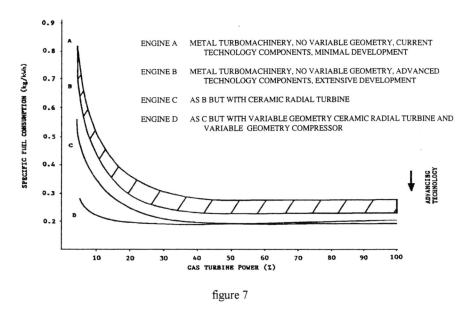

figure 7

TRACTIVE POWER AVAILABLE VERSUS ROAD LOAD REQUIREMENT.

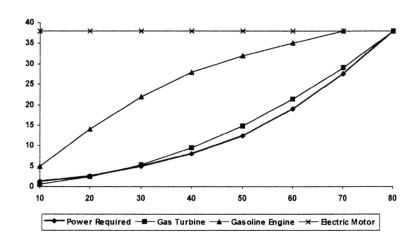

Figure 8

COMBUSTION EMISSIONS: GAS TURBINE v I.C.ENGINES

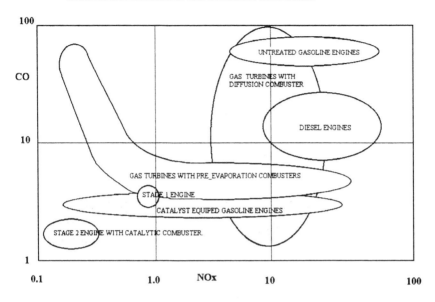

Figure 9.

HYBRID SYSTEMS

Hybrid systems overcome the limitations of a full electric vehicle by generating electricity on board the vehicle to charge batteries or capacitors or to put energy into kinetic energy storage devices such as advanced flywheels. When operating in the rural or motorway environment, the generator drives the vehicle either conventionally or via an electric transmission. The system in effect gains the advantages of both worlds whilst minimising the disadvantages. It does however, carry the cost and weight disadvantages of a battery pack and electric drive-train and will experience catalyst start-up difficulties with conventional internal combustion engines. However, of all the systems available in the near term, the hybrid drive-train comes closest to delivering a solution. It is this system that forms the basis for the two designs discussed below.

VEHICLE 1. A LONDON TAXI WORKING IN A ZERO EMISSIONS AREA.

Although the most demanding in terms of design and packaging, paradoxically it is this vehicle that delivers the nearest to the required emissions performance, simply because it is always operating in an environment which demands a large element of relatively efficient, pure electric vehicle operation. In designing the vehicle a decision had to be made on which engine to use. The gas turbine was selected for this vehicle for reasons associated with the way in which it is used, namely:-

- Because it suffers deep discharge cycles the generator is started only rarely, so the slow start-up of a gas turbine would not be a problem.
- Also, because of the deep discharge cycle, there would be a high rate of charge which suits the generating efficiency characteristics of the gas turbine.
- Quiet stationary generation, was a customer requirement.
- The pricing structure of the taxi and the economics of operation, could support the added cost of the gas turbine in exchange for long term reliability.

An assessment of the system requirements for a London based Hybrid Taxi Cab.

In order that decisions can be made affecting major drive-train componentry, it is necessary to understand the duty cycle under which the hybrid vehicle must operate. It has become clear that when evaluating for instance, the design requirements for a collection and delivery truck versus a taxi, entirely different constraints apply, which drive the design solutions in very different directions. To resolve this issue, a theoretical cycle was constructed which has been tested through observation and by subjecting a number of London taxi drivers to a questionnaire, which used as its basis an hypothetical situation whereby a zero emissions area is created in London, which is bordered to the north by Oxford Street, to the south by the river, to the west by Park Lane and to the east by the A1, see Figure. 10.

This area was chosen for the following reasons:-

- It is principally a commercial and historic area, therefore permanent residents will be at a minimum, thus reducing the need to give concessions to residents' vehicles.

- It is an area attractive to tourists and would be more so, if traffic density were to be reduced.

- Controlling the vehicles that are allowed access to the area, would improve security.

Duty Cycles.

The cycles employed are based upon a hybrid of European test cycles and data gathered by INRETS, the *Institut National de recherché sur les Transports et leur Securite*, see Figure. 11. An additional cycle has been created, also using ECE-15, combined with constant speed running to simulate a journey to Heathrow from the proposed Zero Emissions area in the centre of London, see Figure. 12. The cycles have been assembled in such a way as to take into account changing traffic density during the day, see Figure. 14.

Overlaying the cycles were repeated daily excursions, which have significant effects upon the final design; they averaged out as follows:-

- Tea breaks are taken in the middle of each 4 hour shift and are claimed to be of 10 minutes duration, though observation suggests 15 to 20 minutes. During the day these breaks are taken around half a mile outside the zero emissions area. Drivers stated that they would not object to leaving the engine running during this break, provided of course that noise and vibration levels were tolerable; they usually remain in the cab for the duration of the tea break.

- At night, drivers take their breaks in the area.

- During the day, the average cab operates for around 90% of the time in the proposed zero emissions area. At night, the average cab operates for around 50% of the time in the area.

- During the day, each cab makes between 5 and 6 trips to a railway station outside the area.

- On average, only one trip a week is to Heathrow or Gatwick.

- During an eight hour shift, a cab picks up twenty fares.

- The average shift mileage is 125 miles.

For the purposes of computation, it was assumed that at the end of every 1 hour period of zero emissions operation, the cab remains outside the area for 15 minutes, during which time it is both charging the batteries and operating to the standard city cycle, see Figure. 14.

Vehicle specification.

The data describing the vehicle characteristics and its required performance envelope, is shown in Figure. 16.

The key minimum performance parameters are:-

- An ability to start at Gross Vehicle Weight on a 1 in 3 hill, which determines the motor size.

- An ability to maintain 70 m.p.h. up a 1 in 20 slope on the generator only, which determines the generator size.

- The charge to discharge relationship, which determines the battery pack size and reserve charge capability.

TAXI PROGRAMME RESULTS AND CONCLUSIONS.

The energy flows that occur during a single standard ECE-15 cycle, are shown in Figure. 11. The energy flows for the congested city cycle and the Heathrow cycle are shown in Figures. 12, 13 and 14.

The standard ECE-15 is the most arduous cycle.

Two generators were considered, a 40 kW unit representing the present Volvo ECC unit and a 65 kW unit, thought more appropriate to a vehicle the size and weight of the proposed London taxi.

The available performance range with a 40 kW generator and 75 kW motors is shown below and in Figure. 16.

Performance summary.

The average discharge rate for the taxi whilst operating to the standard ECE-15
= 4.83 kW.

The average discharge rate for the taxi operating to the congested ECE-15
= 2.72 kW.

The average charge rate whilst operating to ECE-15 (40 kW generator) = 33.15 kW.

The average charge rate whilst operating to ECE-15 (65 kW generator) = 56.48 kW.

The minimum usable battery capacity is 10 kWh., i.e. 15kWh nominal capacity.

Assuming no parasitic drain, a 40 kW cab can complete 1.7 hours of running per charge session.

Assuming no parasitic drain, a 65 kW cab can complete 2.9 hours of running per charge session.

A 40 kW cab can maintain 47 m.p.h. up a 1 in 20 slope and has a sustainable max. speed of 76 m.p.h.

A 65 kW cab can maintain 65 m.p.h. up a 1 in 20 slope and has a sustainable max. speed of 92 m.p.h.

The 65 kW generator was chosen.

C52

PROPOSED ZERO EMISSIONS AREA.

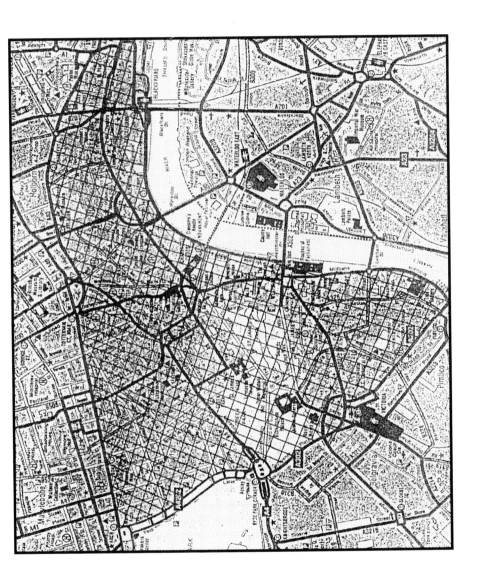

Figure 10

ENERGY REQUIREMENTS TO MEET DERIVED DUTY CYCLES.

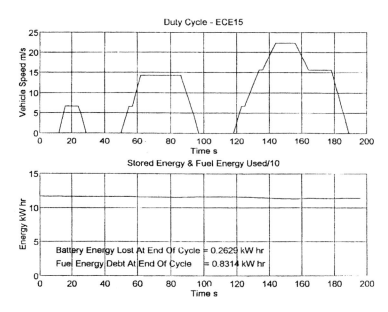

Figure 11

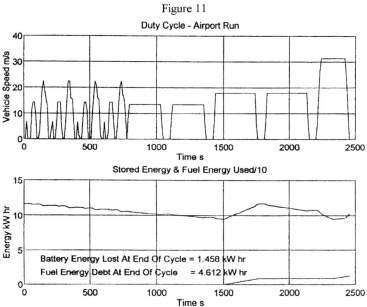

Figure 12.

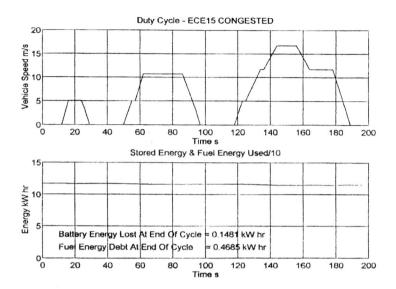

Figure 13.

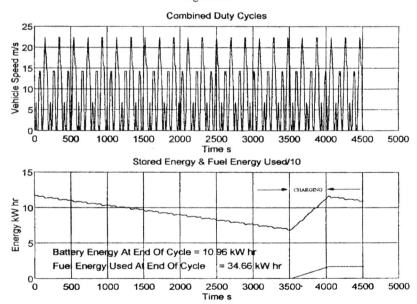

Figure. 14

FULL WORKING DAY DUTY CYCLE.

Phase	Cycle Stage	Description
1	Congested City 1	0700-0900, 90% utilisation. Average speed 14 kph. Based on ECE-15. Total distance 25.2 km.
2	Standard City	0900-1530, 65% utilisation. Average speed 18.7 kph. ECE-15 Total distance 31.5 km.
3	Congested City 2	1530-1830, 75% utilisation. Average speed 14 kph. Based on ECE-15. Total distance 31.5 km .
4&5	Airport Run	Zero emissions area to Heathrow Airport. Average speed 47 kph. Total distance, round trip 64 km. 4 standard ECE-15 cycles. 2 low speed cycles. 2 medium speed cycles. 1 high speed cycle.
	Total Daily Duty	200 km 124.3 miles.

Figure. 15

TAXI CHARACTERISTICS AND PROPOSED PERFORMANCE REQUIREMENTS.

Drive Train Parameters	Assumptions
Engine power envelope.	<20 to 65 kW of efficient generation.
Usable battery storage capacity 100 to 30% discharge.	10 kph.
Motor characteristics.	93% efficient and 300 Nm static torque.
Generator characteristics.	93% efficient 5 to 60 kW.
Gearbox characteristics.	95% efficient over working range.
Power handling characteristics.	95% efficient 0 to 100 kW.
Battery power density.	275 W/Kg over 60 sec's.
Battery charge capability and efficiency.	85% charge efficiency and 15 mins to 90% charge from 30 % charge level.

Vehicle Requirement	Target
Hill start.	1 in 3 and 0.1g at gross vehicle mass.
Max. speed and gradient.	70 mph up 1 in 20.
Min. period between charges.	2 hours on city cycle.
Charge envelope.	40 and 65 kW max. (vehicle stationary). Range:- 20% min. discharge point. 70% max. discharge point.

Vehicle parameter	Value
Gross vehicle mass G.V.M.	2150 Kg.
C.d.	0.36.
A.	2.6 sq.m.
Rolling radius.	0.34 metres.
Tyre losses.	0.012 x G.V.M.

Figure 16.

GENERAL ARRANGEMENT OF THE TAXI.

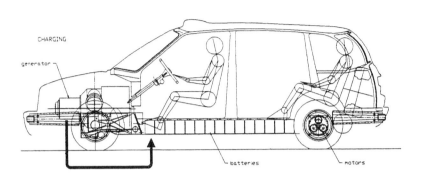

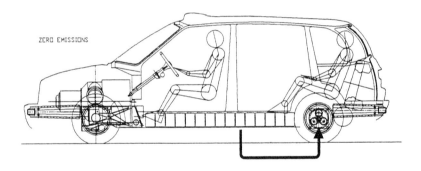

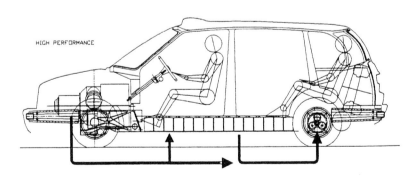

Figure 17.

VEHICLE 2. A SMALL, GENERAL PURPOSE, PRIVATELY OWNED VEHICLE.

The principal assumption in designing this vehicle, was that the purpose of the novel drive-train was to save fuel, rather than to run for extended periods in zero emissions mode. To this end, weight was very important and the battery size was to be limited to that necessary to trickle the vehicle along in traffic, providing only a minimal battery range of around 3 to 5 miles at speeds of less than 20 m.p.h. To achieve maximum battery life from conventional lead acid batteries, only shallow discharge cycles were proposed, with regular engine start-up and low charge rates. This set of requirements did not favour the gas turbine and so a dual phase, direct injection, gasoline or natural gas reciprocating engine was chosen.

Generator design.

Electrical generators are most weight and cost effective when running at high speed; conventionally, low speed engines drive generators through a gearbox to achieve this, but this solution is noisy, occupies a lot of room and removes a lot of the weight advantage that comes from operating the generator at high speed. For this vehicle, this solution was rejected in favour of generating the flux at the periphery of a flywheel, so achieving a high rate of flux cutting without the need for a step-up gearbox. Such systems tend to be less efficient than long cylindrical generators because of the less advantageous active to non-active coil lengths, but it was better than suffering the gear losses from a step-up drive. By choosing a battery pack capacity that would provide sufficient voltage to run the generator as a motor i.e. around 30% of generator maximum voltage, the need for a conventional starter was eliminated. It also provides a quiet start, essential for the modes of operation envisaged.

Hybrid drive systems.

Hybrid drives come in many forms, including using a conventional mechanical drive for the occasions when the vehicle is being used in a non-emission sensitive area, with a near stand-alone system to operate in a zero emission mode. These systems are generally called parallel. Such systems have a great deal of merit and are marginally more efficient than a full electric drive-train, but they do carry the requirement for more drive-train components than are absolutely necessary and can never match the refinement that a full electric drive-train can offer. From an engineering aesthetics point of view, they are not an elegant solution. The full electric transmission (series) route was chosen. The modes of operation of the system can be seen in Figure. 18.

Choosing the electric transmission route freed up a number of options, one of which was engine configuration. The fact that a flywheel generator is longer than the flywheel it replaces, puts the length of a transversely mounted, conventional four cylinder engine under scrutiny. By using a three cylinder engine, that problem was overcome and since there is no mechanical connection between the wheels and the engine there is no need to react the torque from the engine against the body structure, hence engine mounts can be very flexible and the less desirable firing order vibrations of a three cylinder engine can be attenuated. The specification of the vehicle was as follows:-

General specification.

Kerb weight.	650 Kg.
C.d.	0.28
Projected area.	2 sq. m.
Engine.	3 cyl gasoline, 35kW.
Traction motor.	Switched DC, 40kW.
Generator.	Direct drive flywheel.

The emissions performance of this configuration will of course, be entirely dependent upon the way in which the vehicle is used. If it is used solely within the city, it will give up to a 40% improvement in all emissions, including CO_2, over a conventional vehicle. If it is used solely in rural and motorway conditions, it could be slightly worse than a conventional vehicle.

The overall performance and urban cycle economy of the vehicle are remarkably good, see Figures. 19 and 20, returning a predicted gasoline fuel economy of around 80 m.p.g. when operating to the ECE-15 cycle. Overall performance is predicted as follows:-

Maximum speed.	91 m.p.h.	146 k.p.h.
0 to 60 m.p.h.	12 sec's.	
Range on batteries.	3.6 miles	5.8 kilometres
Maximum speed on batteries.	32 m.p.h.	51 k.p.h.
0 to 20 on batteries.	19 sec's.	
Urban fuel consumption.	3.5 litres / 100 km	
56 k.p.h.	4 litres / 100 km	
75 k.p.h.	5.5 litres / 100 km	

It is probable that a vehicle operating to an average, privately owned, duty cycle would see an overall 20% improvement in fuel consumption and emissions reduction. The general arrangement of the vehicle and its operating modes are shown in Figure. 18.

CONCLUSIONS.

The vehicles and the technologies described above are capable of delivering the following improvements in carbon dioxide generation.

Conventional 1996 technology base case.	100%
Average vehicle weights reduced by 30%.	85%
+All vehicles running on fully converted natural gas.	64%
+All vehicles have lightweight hybrid transmissions.	51%
Inner city vehicles 1996 technology.	140%
Inner city vehicles running on fully converted natural gas.	105%
+Inner city vehicles have zero emissions hybrid transmissions.	63%
Intercity heavy transport base case.	100%
All vehicles running on fully converted natural gas.	75%

Since the greater part of the heavy transport vehicle weight is payload, it is assumed that weight reduction of the prime mover will be significantly less effective than would be the case for the motor car.

Comparing these predictions against the target requirements of 62% to meet 1990 level targets and 52% to meet 1990 minus 10% targets, it is clear that a combination of lightweight vehicles with advanced power units, hybrid vehicles and natural gas powered vehicles can reach the first requirement, provided that designs are implemented immediately. However, if the motor vehicle is to achieve a reduction equivalent to its part in a 10% overall reduction of CO_2 by the year 2010, all vehicles would need to be natural gas powered hybrids, but the fleet would still fall short of the target since there is no emerging technology to improve heavy goods vehicles beyond fuelling by natural gas.

The weight reductions implied by the analysis are very substantial and are very much against present trends. It is extremely unlikely that they will happen voluntarily. Furthermore, since those technologies that are being developed which might answer the need, are mostly being developed in America and Japan, and since so few of Britain's cars are now designed in this country, our ability to do something about it is limited. It appears therefore, that the achievement of the 2010 goals will demand draconian legislation, which presently, Europe is reluctant to contemplate and which the U.K could not implement unilaterally.

C527/

GENERAL CONFIGURATION AND OPERATION OF SMALL GENERAL PURPOSE VEHICLE.

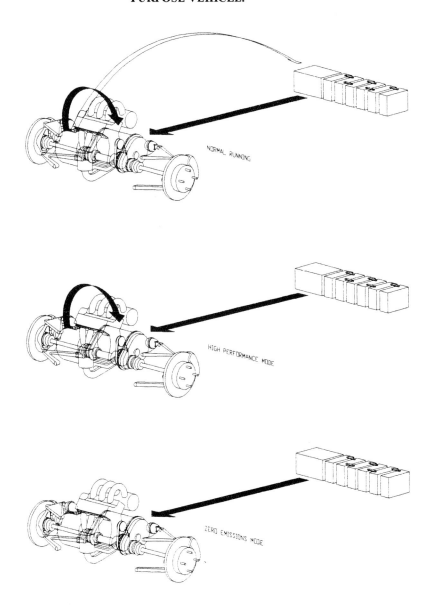

Figure 18.

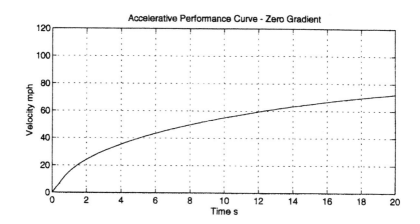

28.76kW Available Power

Maximum velocity 91.66mph

Figure.19

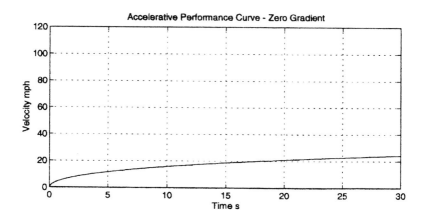

2.465kW Available Power

Maximum velocity 32.41mph

Figure. 20.

C527/

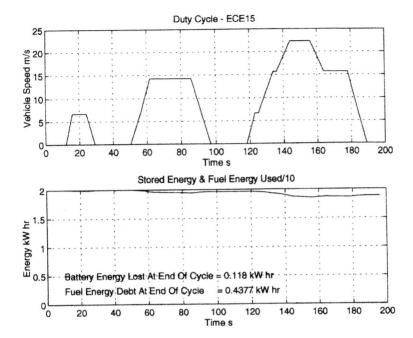

Figure. 21.

REFERENCES.

1. L. Conti, M. Ferrera, R. Garleseo and E.Volpi. *Rationale of Dedicated Low Emitting CNN Cars* SAE paper 932763. 1993

2. RAW Chevis, J Everton and P.P.Walsh. *Hybrid Electric Concepts Using Gas Turbines.* Autotech 1991 Birmingham.

3. K.R.Pullen, *A Case for the Gas Turbine Series Hybrid Vehicle.* Autotech 1991. Birmingham.

4. T.M. Sawyer. *An Analysis of the Operational Characteristics of Advanced Hybrid Vehicle Systems.* PhD thesis 1995.

5. J.N.Randle, MR Jones, TM Sawyer. *An analysis of Future Power Train Requirements.* British Association Lecture 1993.

C527/

l transport for cities

MP FEng
ical Advisor, GEC ALSTHOM, UK

SYNOPSIS

In the first four post-war decades there were many technical innovations in the railway industry but this did not stop it losing market share to road transport. In the past decade rail has started to hold its own and, in the new millennium, environmental issues will necessitate major growth in the volumes of passengers and freight carried by rail. However, in the new environment, taxpayers are not prepared to write a blank cheque for new infrastructure. The challenge to the industry will be to find ways of attracting more passengers, reducing energy consumption, reducing both the first cost and the operating cost of equipment and making greater use of the existing railway assets.

THE ENVIRONMENTAL CHALLENGE

Growth in transport demand

A recent Royal Commission report[1] presented the growth in road transport over the past 40 years. The data for passenger cars are shown in Figure 1. The line of best fit shows a steady increase of 12.5 billion passenger kilometres per year over that period. In contrast, Britain's rail network provides around 50 billion passenger kilometres per year. A comparison of these figures show the enormity of the task if the growth in transport is to take place on the railways rather than on the roads. For the railways to absorb *all* the increase in passenger transport would require a growth rate of 25% per annum.

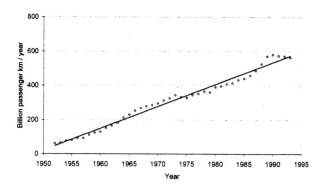

Figure 1 Growth in personal mobility by road

Energy use by transport

In the 1990s transport accounted for more than one third of Britain's energy consumption; most of this was liquid fuels used in road transport. In 1960 rail used 32% of the primary energy taken by transport[2] but, with the change from steam to diesel and electric power and the massive growth in road transport, rail now accounts for less than 5%. In fact the electricity consumed by rail transport is less than that used by road lighting.

Responses to pollution caused by transport

The Houghton Report identified transport as the major cause of pollution in Britain. Poor air quality, noise, visual intrusion as well as global warming are directly related to the growth in transport. All political parties now recognise that the increase in road transport cannot be allowed to proceed unchecked and, for several years, there has been an informal policy that new road construction has been allowed to lag the growth in traffic. A result of this is that, on a 1996 bank holiday, 300 km of traffic jams were recorded on Britain's network of motorways and trunk roads[3]. This is a policy that can only be used for so long without an unacceptable degradation of the quality of life, but experience in gridlocked and polluted cities, such as Bangkok, has shown the lengths to which people will go to avoid being separated from their cars.

How these environmental factors will affect rail transport in the next millennium is difficult to assess. It seems inevitable that the demise of the *Great Car Economy* will lead to pressure for a major expansion of rail services but the industry will have to demonstrate that this can be achieved with good energy efficiency and without detriment to the environment. The saga of the Channel Tunnel Rail Link has shown that local communities do not welcome new railway lines in their "back yard" any more than they welcome motorways or airport runways.

INCREASING THE USE OF THE RAIL NETWORK

Squeezing more traffic onto existing routes

In comparison with the motorway network, the railway is under-utilised. Drivers on a motorway regularly drive with headways of 1 to 2 seconds and, consciously or subconsciously, rely on looking, not just at the car in front of them, but at what is happening several vehicles ahead.

On a railway the situation is different and, on all Britain's InterCity lines, block signalling is used. This works on the basis that the railway is divided into sections, typically 1 km in length, with an occupied section protected by a red signal at its entry which the driver of a following train is not allowed to pass. The entry to the previous section is indicated by a yellow signal, instructing the driver of a following train to slow down, and so on.

The block system has been developed over the lifetime of the Institution of Mechanical Engineers and is an extremely safe way of operating trains. The disadvantage is that the track utilisation is low. Trains operating at 200 km/h are generally constrained to maintain a separation of around 15 km (a headway of 4 min). As each train is 250m long this means that less than 2 % of the track has a train on it at any given time. (In comparison a motorway full of cars travelling at 80 km/h, each driving 2 sec behind the car in front, achieves c.10 %.)

Communications-based signalling

To increase the throughput of a railway system, communications-based signalling (CBS) can be adopted. One such system is SACEM that has been in use on the RER (high speed metro) in Paris for the past ten years and has recently been installed in Hong Kong, Santiago and Toronto.

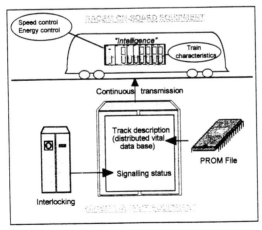

Figure 2   SACEM system

The principle of the SACEM system is that intelligence is distributed between the train and the wayside. The SACEM wayside computer knows the route and transfers this information to the train. It also receives data from the local interlocking to identify where the various trains are on the system. In the Paris RER application there are a number of *virtual blocks* established in each physical track circuit block which enables trains fitted with the right equipment to operate at closer headways than would be possible with the normal block signalling. Cab indications advise the driver of the appropriate speed and apply an emergency brake if a safe stopping trajectory is exceeded. CBS systems can also use a *moving block* architecture in which each train "drags behind itself" a safety zone, dependent on its speed, in which the following train is not allowed to enter.

Signalling in the next millennium

CBS systems can eliminate excessive safety zones between trains and thus can allow them to travel closer together. Metros using CBS can operate with less than 90 sec headways between trains and 300 km/h high speed trains with 3 min headways. However all these systems work on the traditional basis that a sufficient gap must be left behind a train so that, if it stops dead, the following train can stop safely.

If a dramatic increase is to be achieved in the use of the infrastructure, it will be necessary to question this fundamental principle of railway operation. Realistically a train cannot stop dead and thus, if adequate information can be provided to the following train, the gap between trains is only necessary (at least in theory) to cater for the reaction time of the equipment and the variability of the brakes between the two trains. In the canons of railway signalling this counts as heresy. The challenge for engineers in the next millennium will be to find a way of implementing such an heretical system that is not detrimental to the enviable reputation for safety built up over the past century by the present signalling philosophy.

ENERGY CONSUMPTION

Aerodynamic drag

A train has a lower rolling resistance than road vehicles and most of the tractive effort goes to overcome aerodynamic drag. There has been a marked improvement in the drag of trains. This has been achieved by the aerodynamic design of the nose and tail and also by careful attention to the roughness of the surface, eliminating protruding doors, window frames, ventilators and guttering, minimising the gap between coaches, covering below-floor equipment and a host of small changes. Figure 3 is a comparison of the train resistance curve for the standard British Rail 475 ton passenger train[4] 25 years ago with the latest double-deck *TGV-Duplex* operating on French Railways (SNCF) today[5]. The TGV has a larger seating capacity than the older train and it is 25% lighter despite being designed for twice the speed. The most dramatic change is that drag, and hence energy consumption, at 140 km/h has been halved.

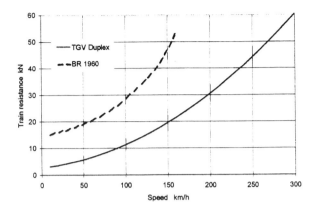

Figure 3　Comparison of train resistance

Rolling stock and equipment mass

There is always a tendency for equipment to become heavier. The easy solution to improving the crashworthiness of a vehicle is to add material; improved passenger facilities, such as powered doors, retention toilets and air conditioning add weight. There is thus a continuing challenge to vehicle and equipment designers to reduce the mass both of vehicle structures and of equipment. An example of the progress in the latter area is shown by the transformer designs in the following table[6]. Lighter equipment requires a lighter bodyshell, allows lower axleloads and thus causes less track wear.

	BR Class 90	BR Class 91	SNCF TGV-PSE	SNCF TGV-A	Eurostar
Rating kVA	6100	7840	4360	4050	7500
Weight, core + copper kg	3300	3960	2340	2010	2470
Ratio kg/kVA	0.54	0.51	0.54	0.50	0.33

Figure 4　Reduction in mass of locomotive transformers

High speed passenger vehicle structural design is frequently dictated by stiffness rather than by strength. This is because of the need to avoid body resonances from amplifying the bogie kinematic frequencies. Body designs using aluminium alloy extrusions concentrate the material where it can be of greatest use to stiffen the structure.

A number of prototype vehicle bodies have been made from composite materials, including wound carbon fibre. Engineers in the next millennium can expect to see a much greater use of non-metallic structures, in the "difficult" structures of bogies and axles as well as in vehicle bodyshells.

Regenerative braking

Most of the energy consumed during acceleration of a train is converted into kinetic energy that is later converted into waste heat when the train is braked. For a 500 tonne train travelling at 300 km/h, $\frac{1}{2}mv^2 = 1.7$ GJ or about 0.5 MWh. Apart from the cost of the energy, the wear on brake pads and, in tunnels or covered stations, the heat coming from the brake disks and rheostats are undesirable side effects of this waste of energy.

Hong Kong MTRC, which runs an intensive service of high-performance underground trains, have calculated that the energy saving of using GEC ALSTHOM regenerative choppers, as opposed to non-regenerative camshaft control, is 53 % of which 10 % is due to eliminating resistor losses and the rest due to regenerative braking. In tests, including the effects of non-receptivity, it was found that the measured saving was 46 %[7]. The economic benefits of moving to chopper control were so great that HKMTRC found it worthwhile to replace camshaft controlled equipment that was far from being life-expired and the replacement paid for itself in very few years.

For a.c. supplied railways forced commutated converters are used. A fleet of multiple unit trains operating on the Kwachon line in the suburbs of Seoul, Korea, uses this technology and most of the braking energy is returned to the overhead 25 kV 60 Hz supply. Apart from major energy savings the new semiconductor controlled systems cost substantially less than electro-mechanical systems of earlier generations.

NEW URBAN RAILWAYS

Mass transit systems

The design of an urban transport system has to be closely linked with the characteristics of the urban area which it serves. Dr. J. Kenworthy has analysed a number of cities[8], 15 in the USA, 11 in Europe and 3 in Asia. Figure 5 shows a comparison of housing and employment density. The Asian cities used in the analysis were Tokyo, Singapore and Hong Kong, all wealthy urban areas - a status to which many developing cities, such as Bangkok, Jakarta, Kuala Lumpur, Manila, Surabaya and Seoul will aspire in the next millennium.

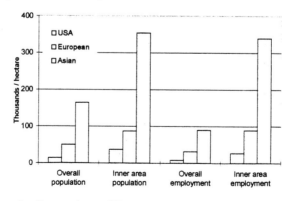

Figure 5   Comparison of North American, European and Asian cities

It can be seen from Kenworthy's figures that the Asian cities have densities of population and employment roughly ten times those in the United States or three times the European average. Public transport is essential in these conditions to maintain an acceptable quality of life; a US-style dependence on the car would create traffic chaos and appalling local environmental damage.

Because of the density of population and jobs, Asian transport systems are of very high capacity. One line of the Hong Kong MTRC carries 70,000 passengers/hour/direction which is more than twice the capacity of the most heavily loaded lines on the London Underground. Seoul, which has seen a ten-fold growth in population over two decades, commissioned its first metro line in 1982 and now has eight lines in operation with more planned. Trains have a capacity of 2000 passengers and operate every 2 minutes. The new millennium will see a continuation of this growth in metro systems as developing cities attempt to escape from the stranglehold of the car.

Suburban rail systems and airport links

The needs in Europe are different to those in the Far East. City populations are not expanding as in Asia - in fact some cities have seen a reduction in the inner area. In many countries the quality of life of many commuters would be vastly improved by better connections between the cities, where people work, and neighbouring towns and outer suburbs where they live. Services such as the RER in Paris or Thameslink in London have made a major difference to many commuters and other services are being planned.

The Heathrow Express, connecting the airport with London's Paddington rail station, is due to be fully open in the summer of 1998[9] and will provide a welcome improvement in the quality of life for many airport users. This is seen as the first of several rail services linking to major airports including through services from Guildford via Heathrow to Watford and a service from Heathrow to Kings Cross/St Pancras and thence south of the river.[10]

Light rail systems

In smaller cities and towns an expensive high-capacity transit system operating in a single corridor is often not the best solution to the transport needs of a dispersed population employed over a wide area. A network of light rail routes achieves a far better penetration of residential areas and allows transport to the heart of the business district rather than to a station on the edge. Figure 6 shows the growth in ridership for the first years on the Manchester Metrolink light rail system. The 1991 figures relate to the ridership while the route was operated as a BR suburban service, 1992 was disrupted by construction and 1993 was the first full year of LRT operation. By 1995 twice as many passengers were carried as when the route was a conventional suburban service and farebox revenues adequately cover operating costs. Work is now in hand to extend the successful Metrolink network.

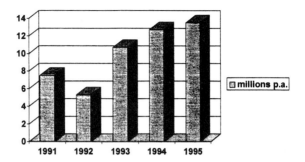

Figure 6   Growth in Metrolink ridership

CONCLUSIONS

Modern rail transport can vastly improve the quality of life both for city dwellers and for long-distance travellers. The further development of intelligent control systems will be able to increase the utilisation of railway systems and thus reduce unit costs to the operator, the fare to the passenger or the subsidy from the taxpayer. In the next millennium we can expect to see major improvements in energy consumption, due to lighter structures, better aerodynamics and regenerative braking, as well as increased reliability due to the replacement of electro-mechanical by solid-state systems.

In rapidly growing Asian and South-American cities, it is probable that the construction of high-capacity metro systems will continue. However, in most European cities we are likely to see a different picture with a growth in air-rail links, further integration and penetration of suburban rail services into, and through, city centres and the establishment of networks of light-rail systems.

Many urban areas across the world are on the verge of a transport crisis; for a few the crisis has already arrived. Pollution and congestion are endemic and transport consumes a large part of national energy supplies. If cities are not to seize-up early in the new millennium, new public transport will be needed. Modern urban rail systems offer an effective way of maintaining personal mobility, which is key to contemporary life, while protecting the city environment. Mechanical Engineers will continue to play a large part in their development.

REFERENCES

[1]   Transport and the Environment, Royal Commission on Environmental Pollution, HMSO 1994.

[2]   Energy conservation in the UK, HMSO 1974

[3]   The Guardian, 28 March 1997

[4]   Proceedings of British Railways Electrification Conference, London, 1960 pp 75, 89, 119, 144, etc.

[5]   P Delfosse, Les TGV-Réseau et le Duplex. Revue générale des chemins de fer, Nov-Dec 1996

[6]   B Cousins, P C Bennett, A new generation of traction transformers, GEC ALSTHOM Tech. Rev. No. 8, 1992

[7]   A. D. Moyes, Hong Kong Metro's traction equipment: the drive for efficiency, Proc. IMechE, Part F, 1992.

[8]   J. Kenworthy, Developing sustainable urban transport in Asian cities, Asia Urban Rail Congress, Singapore, 1996

[9]   R Hoare, The Heathrow Express - a new opportunity, Air-Rail Conference, London, November 1996

[10]   London Airports Surface Access Study (LASAS), September 1996

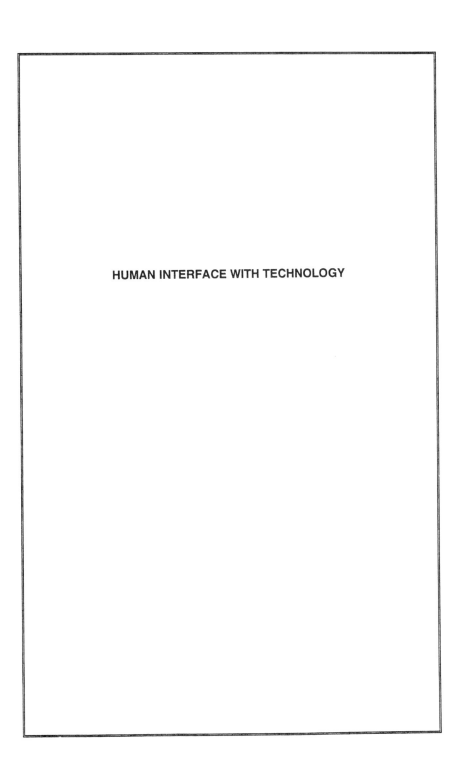

HUMAN INTERFACE WITH TECHNOLOGY

Minimally invasive surgery: the surgeon's interface

McCLOY BSc, MD, FRCS
or Lecturer and Honorary Consultant Surgeon, University Department of Surgery, Manchester Royal Infirmary,
chester, UK
fessor R STONE BSc, MSc, CPsychol, AFBPsS, MErgS, EurErg, FRSA
ctor and General Manager, VR Solutions Limited, Salford, UK

ABSTRACT

Surgeons performing video-assisted 'keyhole' surgery have faced the challenges of losing
tactile sense from their fingers, operating on three-dimensional organs whilst looking up at a
two dimensional monitor, and viewing magnified anatomy in unusual perspectives. Surgeons
are now seeking a new interface with the patient via novel instrumentation and imaging
techniques. Prototypes of haptic force feedback simulation and three dimensional imaging are
being tested. Virtual reality simulators have been developed to aid training but have had to be
scaled down from complex graphical representations of organs to a basic skills trainer (MIST
VR).

BACKGROUND TO THE INTERFACE

The minimally invasive surgical revolution

The last decade, triggered by the development of video-chip miniature camera technology, has
seen an explosion in 'keyhole' surgery in many surgical fields, with the best known operation,
being in general surgery, laparoscopic cholecystectomy (removal of the gallbladder for
gallstone disease). Surgeons have faced the challenge of losing the tactile sense from their
fingers and operating on 3-Dimensional organs whilst looking down at their patient at open
surgery. Instead, they have to look up to a 2-Dimensional TV monitor, displaying a
magnified view of an anatomy in unusual perspectives. In short, keyhole surgery uses hardly
any of the skills most senior surgeons have spent a professional life-time striving to acquire.

Keyhole surgery has become a patient-led revolution, with the obvious advantages of minimal scarring, less pain, shorter hospitalisation and rapid return to normal activities and employment. However, surgeons have found it difficult to keep pace with the change. The majority have risen to the challenge of retraining with enthusiasm and responsibility, but a minority have been driven by financial expediency and miss-placed bravado into attempting these new procedures with reckless abandon, forgetting the principles of open surgery, which still apply whichever way the operation is undertaken. Not surprisingly, they have become the subjects of litigation and attracted the attention of the press and arguably threaten to hold back the most significant surgical advance since the development of general anaesthesia.

The issues of training and evaluation of the new instrumentation are being tackled by the Royal Colleges of Surgery, in collaboration with specialist groups such as the Association of Endoscopic Surgeons of Great Britain and Ireland (AESGBI), and have attracted specific funding from the Department of Health (DoH) for training centres - including the North of England Wolfson Centre for Minimally Invasive Therapy, based at Manchester Royal Infirmary.

In the UK, unlike the US or some countries within Continental Europe, live animal-based training is prohibited. As a result, primary laparoscopic experience is typically fostered through the remote handling of artificial 'tissues', sugar cubes, grapes, raw chicken tissue, plastic tubing and the like in a basic 'box' simulator (Figure 1, equipment on right-hand side).

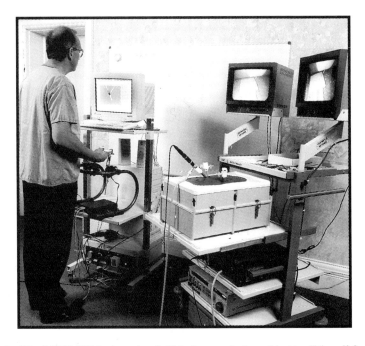

Figure 1. The MIST *VR* trainer (on left) being used alongside 'traditional' basic 'box' laparoscopic surgical skills training trolley (on right).

C527/C

This situation, coupled with increasing medico-legal and certification pressures in the UK and the clear requirement for objective assessment of trainees, means that the provision of even a *basic* fidelity virtual human simulator is seen as instrumental to the development of specific surgical skills. The surgeons' new interface will be with a virtual world.

Medical application of Virtual Reality technologies
Medical applications of Virtual Reality (VR), whether for anatomical education, simulation of physiological processes, diagnosis or surgical training, place considerable demands on VR technology in this field. When dealing with structured virtual worlds - such as engineering models, architectural plans - building virtual simulations can be reasonable straightforward. However, the unstructured and highly dynamic nature of human tissues undoubtedly presents VR researchers with their greatest challenge to date.

R&D at the Wolfson Centre, funded by the Department of Health (DoH) in conjunction with the Wolfson Foundation (1994-96), has concentrated on integrating the skills and talents of a number of academic and small company teams throughout the UK, together with background "know-how" from the Virtual Reality & Simulation (*VRS*) Team at VR Solutions Limited, Salford (1-3).

Initial work on the project, 1994-95, involved the development of the main components of a 'full-blown' laparoscopic virtual reality simulator, involving deformable tissues, realistic tissue rendering, force feedback instrumentation and spectacle-free auto-stereoscopic 3-D display (3).

Unlike artificial, or synthetic body trainers (with relatively costly replacement 'parts'), VR has the potential to reproduce the natural characteristics of human anatomy, requiring only a re-setting of the host computer, once a particular training session has been completed. The use of VR techniques will undoubtedly compliment artificial anatomy by providing the 'flight simulator' qualities that some surgeons are demanding - the ability to generate cost-effective *"what if"* scenarios, such as progressive skill levels, anatomical ambiguities (abnormal anatomy and excess fat), bleeding and instrument failure.

DEVELOPMENT OF A VIRTUAL REALITY SURGICAL SIMULATOR

A number of key developments gave the DoH/Wolfson collaborators confidence that prototype full- and part-task British simulators were achievable.

Theatre task analysis
The first stage of the project, 1994, involved in-theatre observation and recording sessions at Manchester Royal Infirmary, using video and digital laparoscopic surgical image storage techniques. The aim of this simple task analysis was to obtain a clear understanding of the performance and ergonomic features of the surgeon's task and to evaluate a number of different forms of media for image capture (and subsequent digitising for anatomical texture mapping). These exercises, together with lengthy briefing sessions from practising and trainee surgeons produced the following main conclusions. 1) Any form of head-mounted display, be

it for full or partial immersion, would not be acceptable to British surgeons, for image quality, in-theatre communication and awareness of surroundings, or ergonomic reasons. 2) There was approximately a 50:50 split in opinions as to the need and value of stereoscopic viewing. 3) There was a similar split in opinions regarding the importance of tactile and force feedback. Some surgeons claimed that vision was the overriding sense; others insisted that a simulator would have to include a reasonable representation of such features as elasticity, instrument constraints within the body cavity and varying degrees of movement "viscosity". 4) Surgeons wanted the VR simulator set-up to be as close to that used in the operating theatre, thereby enhancing the ergonomic aspects of transfer of training. 5) The fidelity of the simulation need not be totally representative of the real procedure. 6) Texture acquisition, for subsequent VR images, proved difficult during *in vivo* laparoscopic or open surgical intervention (and not of a high enough quality).

Simulator Display system
Initially, the *VRS* Team chose to adopt a display technology which did not require any form of head- or face-mounted peripheral to achieve a stereoscopic effect. Richmond Holographics, based in London, developed an autostereoscopic display system that used conventional LCDs combined with Holographic Optical Elements (HOEs). HOEs are devices made by recording the interference pattern which results when two or more laser beams intersect. Unlike ordinary holograms they do not contain an image. Instead, they are used to bend, focus and otherwise modulate light in unusual and frequently complex ways with properties that cannot be achieved using conventional optics. One HOE can have many different focal points in many different places and can act as a positive and a negative lens, mirror and prism all at the same time. In the Richmond Holographics' autostereoscopic display HOEs make it possible to fabricate a display which directs light from one image into one of the viewer's eyes and light from a second image into the other eye. The HOE allows the direction of the light to be moved without recourse to moving parts, thereby permitting the display to compensate for viewer movement. If the two images are a stereoscopic pair then binocular fusion occurs to produce a three dimensional whole image.

From a human factors point of view, experimental support for the benefits of stereoscopic viewing in remote systems, be they for nuclear decommissioning or even surgery have, to date, been contradictory (4,5). This trend will probably continue until researchers realise the importance of an integrated perceptual-motor approach to designing human-system interfaces for remote inspection or handling systems, rather than assuming that technologies such as stereoscopic viewing will, in isolation, bring significant performance benefits (6).

Deformable tissue software
Medical and computer science students from both Edinburgh and Salford Universities have been involved with the Wolfson Project, specifically in the development of a suite of algorithms to simulate virtual human tissue, featuring real-time, local geometric deformation (at the point of instrument grasp) and force propagation through the virtual 'organ'.

The code has been designed in such a way that any anatomical model may be modified to demonstrate deformable properties. The model is defined by means of nodes, links, and triangles. Nodes may be internal or external. External nodes are points defined on the surface of the model. Internal nodes are used to provide the dynamic links necessary to simulate deformation and force propagation. Each node is assigned a position in space and an individual mass and may be 'fixed'. A link is defined as the connection between two nodes and may be internal or external. Each link has its own spring constant and has three values associated with it: a slip threshold, defining how far a link may be stretched before it exhibits plastic qualities and remains stretched; a split threshold, defining what force must be applied to the link before it splits into two, and a dashpot value, providing a measure of viscous friction. Triangles are formed from three closed links. These are not used in defining how the model will deform, but define the facets of the model on which texture detail is laid. The simulation currently supports only one texture which is defined in a standard SVT format (Figure 2). The input to the simulation is a position given in Cartesian co-ordinates (x,y,z) via a serial port. This causes a node - previously defined as a 'prodnode' - to move a relative distance.

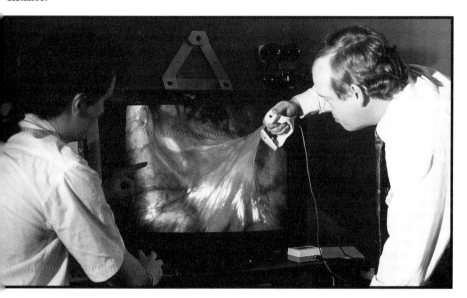

Figure 2. Demonstration of 'deformable tissue' software

Haptic feedback demonstrator

At the outset of the Wolfson/ DoH Project, off-the-shelf haptic feedback systems were not available. A proof-of-concept demonstrator was developed. The general mechanical layout of the interface is based on three, mutually orthogonal actuator rods and by providing the appropriate force to each of the rods, any required force can be applied to the tip. The rod forces are supplied via cranks fitted to servomotor shafts. Each motor shaft is fitted with a rotary position sensor and the actuator rods are fitted with force sensors close to instrument tip. Rotary potentiometers and optical incremental shaft encoders are used to sense the motor shaft positions. The former are used in conjunction with a commercial 12-bit analogue-to-digital converter card in a standard PC. This arrangement gives 4096 position steps within the 600 shaft movement but the potentiometers exhibit problems with noise, non-linearity and conversion time. The shaft encoders are used in conjunction with a PC interface card (in-house design and build) to give 20,000 steps per revolution or 3,333 easily readable, noise-free steps for the 60 degrees extent of rotation. This level of resolution is required for smooth application of force, viscous damping and mass simulation. The force sensors are based on U-shaped steel springs, which suffer bending moments as they transmit a force, such that the foil strain gauges which are attached at the point of maximum strain but on opposite faces suffer equal but opposite strain. When used as a half Whetstone bridge the gauges produce about 0.3mV per Newton. The Wolfson Team is grateful to the Department of Mechanical Engineering, Design and Manufacture at the Manchester Metropolitan University for supplying the springs and providing instrumentation advice.

If a simple run-time graphical model is wholly hosted on the PC, then, by interleaving parts of the graphics rendering with the force control, it is possible to produce a basic demonstration. However, when a complicated mechanical model of deformable human tissue is being run on a locally networked VR computer (Silicon Graphics Onyx), this approach is not possible. It is believed that this can be overcome by transmitting a simplified local force model from the VR computer at video-frame rate to the controlling PC, allowing the PC to make the final calculations. The parameters which could be passed are force, first and second derivatives of force with respect to distance, mass, and damping factors. The demonstrator has been developed to feature "pulsating" effects and tissue "grasp and tear".

MIST *VR* - a part-task skills assessor and trainer
Starting in 1995, to overcome many of the developmental problems discussed above, MIST *VR* (Minimally Invasive Surgical Trainer - Virtual Reality) was designed by the Wolfson/VR Solutions Team, in collaboration with Virtual Presence Limited of London, as a part-task surgical skills trainer to provide a cost effective means of fostering basic perceptual-motor competencies for the "remote" handling of human tissue in minimally invasive procedures (3). MIST *VR* provides a means by which these skills can be practised, recorded and, most importantly, quantified.

The needs of British surgeons dictated the nature of the MIST *VR* simulator - simplicity, affordability and recordable trainee performance. Furthermore, many of the tasks benefit from the additional experience of *VRS* Team members in the related field of human performance assessment for robotics and telepresence and many of the underlying performance analysis algorithms are based on established results from applied psychological experiments. The system comprises of a standard 200 MHz Pentium PC with 32 Mb RAM. This is linked to a jig containing two laparoscopic instruments held in position-sensing gimbals with 6 degrees of freedom (Figure 1, MST *VR* is being used on the left). This provides real-time translation of the instrument movements to the graphic display. An accurately scaled operating volume of 10cm^3 is represented by a 3-D cube on the computer screen. The image zoom and size of target objects can be varied for different skill levels. Targets appear randomly within the operating volume according to the skill task and can be "grasped" and "manipulated" with the virtual instruments (Figure 3).

MIST *VR* has tutorial, training, examination, analysis and configuration modes. Six tasks have been designed to simulate some of the basic surgical manoeuvres performed during laparoscopic surgery. During the tasks, accuracy, errors and time to completion are logged in a spreadsheet. After completion, data analyses permit overall and individual task performance as well as right or left handedness to be quantified and displayed. Comparisons and statistical evaluations can be made between training sessions and trainees.

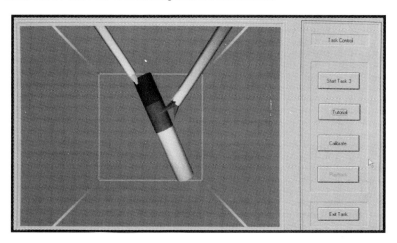

Figure 3. A view of the monitor screen during one of the MIST *VR* skills where the stick-like object is being passed from instrument to instrument (right/left hands in sequence) with the help of colour cues (blue for successful grasp, green for next segment to be grasped), practicing the surgical technique of "walking" down tissues

The applications of MIST *VR* will be primarily as an assessment or examination tool rather than a pure laparoscopic trainer. For the first time quantification of performance allows for aptitude testing and could even assess competence to practice. Surgeons may find this last aspect contentious and worry that it may have parallel implications to flight simulators and airline pilots!

The main advantages of MIST *VR* are that it is now commercially available running on a high-end PC. It is based on abstracted graphics so that surgeons are not distracted by the appearance of 'virtual organs' which are not true-to-life. This also makes the system multi-disciplinary and relatively simple. The quantification of performance provides a powerful new tool for surgical training.

References

1 . Stone, R.J. A Year in the Life of British Virtual Reality: Will the UK's Answer to Al Gore *Please* Stand Up?! *Virtual Reality World* 1994; **2**(1).

2. Stone, R.J., Connell, A.P. The UK's Virtual Reality and Simulation Initiative: One Year Later. *Virtual Reality World* 1994; **2**(5).

3. Stone R. J., McCloy R.F. Virtual Environment Training Systems for Laparoscopic Surgery; Activities at the UK's Wolfson Centre for Minimally Invasive Therapy. *The Journal of Medicine and Virtual Reality* 1996; **1** (2):42-51.

4. Yorchak, J.P. Teleoperator Human Factors Study. MCR 86558; Martin Marietta, Denver, CO; May, 1986

5. Sheridan, TB. Telerobotics. Plenary Presentation for *10th IFAC World Congress on Automatic Control*; Munich, July, 1987.

6. Stone, R.J. Future Trends in MMI Design for Telerobotics in Hazardous Environments. *CERT Discussion Meeting: "Designing the Human-Machine Loop"*; Toulouse, June 1988.

C527/0

he human dimension in future virtual design/prototyping /stems

fessor R S KALAWSKY PhD, MSc, BSc, CEng, MIEE, FRSA
fessor of Human-Computer Integration, Advanced VR Research Centre, Loughborough University, UK

ABSTRACT

The rapid pace in technological development has created the need for, and development of, powerful computer based design tools. These tools are now radically changing the way manufacturing industries design their products. Later generation systems enable the designer to see and interact with a complete synthetic representation using virtual prototyping and virtual manufacturing concepts. The emergence of virtual reality based systems enables the designer to work in a genuine 3D environment, which has many attributes of the real world - though not all. Unfortunately, problems are created when the user interacts with the system but these problems mainly tend to be associated with the complex human factors issues. Therefore, early identification of the human factors issues is paramount. This paper looks at the important human dimension in future virtual design/prototyping systems.

1. INTRODUCTION

Interaction with a computer involves a complex system known as a 'human operator'. Unfortunately, unlike the computer, our understanding of the human operator is less than complete and the desired results of human-computer interaction are not always as expected. These can manifest themselves as poor user performance through to long-term side effects (which do not always show up immediately) such as repetitive strain injury. In fact, at first sight there are so many inter-related human factor issues that they seem to be very daunting and impossible to resolve, but these factors MUST be optimised to ensure that the end user can perform the required task efficiently and effectively.

2. HUMAN FACTORS

Human factors is the science by which we determine the way a user interacts with a system whether it be a mechanical device or a complex embedded computer system. Human physical and cognitive capabilities have a finite limit and beyond this, unpredictable response can occur.

One of the difficult aspects of human factors is its multifaceted nature. For instance, when a 'human factor' breaks down (human error occurs) in an aeroplane cockpit, we are only too aware of the result. Inappropriate human factors during the design stage of the cockpit can lead to unacceptable pilot workload which in turn leads to a disaster. Although human factor issues are addressed in the aircraft industry, in designs involving lesser forms of user interaction human factors are frequently ignored. Although inappropriate human factors at this level may not be life threatening, a lack of attention to human factor issues can significantly affect the quality of the user interface, thus making it very difficult to use a particular system. Moreover, the user's productivity can be seriously impaired.

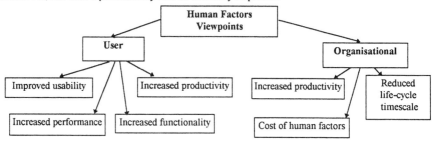

Figure 1 Two important human factors viewpoints

3. USER CENTERED ISSUES

When dealing with the human factors issues of a computer aided design tool it is necessary to think in terms of the person who is going to use the tool.

3.1 Usability

Usability is a multi-dimensional term that relates to different aspects of a product or system and involves consideration of factors such as ease of use, flexibility, error handling, provision of help, etc. Shackel, 1981 (1) stated that: *'The usability of a computer is measured by how easily and how effectively the computer can be used by a specific set of users, given particular kinds of support, to carry out a fixed set of tasks, in a defined set of environments'*. Usability engineering has been introduced as a concept that refers to the systematic process for making a product or system more usable in terms of the friendliness of the user interface, the productivity that can be achieved and the ease with which the system can be used. Preece, 1994 (2) refers to: *'a process whereby the usability of a product is specified quantitatively, and in advance. Then as the product is built it can be demonstrated that it does or does not reach the required levels of usability'*.

Usability issues should be evaluated during the early prototype stages of a design because they can greatly influence the final realisation of the product. One of the key features of virtual design environments is their inherent 3D nature. This means that appropriate user interaction must be provided so that users can operate in a genuine 3D spatial environment. Unfortunately, interaction devices such as the conventional mouse are not suited to 3D interaction because primarily it is a 2D interaction device. However, the trend for 3D visualisation/interaction design tools is on the increase which means that there is an urgent need for suitable interfaces. The display technology used to deliver the 3D environment is also evolving rapidly. The original concept of only being able to use fully immersive (head mounted display based) VR systems is making way for alternative approaches. One of the most exciting of these being the use of large format or panoramic display systems. These provide a semi immersive experience by means of a wide field of display system produced by a high resolution TV projection system. With the right hardware it is even possible to project a stereographic image that has a realistic spatial representation. Clearly, this technology is more expensive than conventional desk-top workstation displays and economic pressures mean that it is even more important to be able to quantify the benefits that new technologies bring. In determining the productivity gains that can be achieved with these new tools it is important to consider the human dimension in terms of user performance.

3.2 User performance (productivity, learning, creativity)

Whenever, we think in human factor terms we have to deal with parameters that are highly dependant on each other, the task and the user's capability. Many parameters are based on empirical science and can be precisely quantified (such as the resolving power of the human eye) but these become difficult to deal with when considered in the context of a whole system. To make matters worse some parameters cannot be expressed in this way. Some tend to be very subjective and can only be quantified after sampling a large user base.

In any product involving a user it is necessary to determine the effectiveness of the interface in terms of usability. Where the product is expected to enhance productivity then we must be able to assess the user's productivity in the context of use. Complicated interfaces should be designed to be easy to learn and incorporate intuitive aspects.

3.3 Differing user requirements

Users of virtual design tools will range from computer novices through to technical experts and all will have varied abilities. However, whatever category the user falls into he/she will want to use the design system in their own particular way and therefore each way will be different. Obviously, each will expect the system to cater for their own need. This means that the system provider must be able to interpret the needs of all the different user communities. Understanding these requirements is the job of the human factors specialist who will try to balance the requirements of the task with the functionality provided by the system. The human factors specialist needs to look much further into the way design tools are evolving in regard to the future use of them. The use of virtual reality concepts in the design process will require a careful consideration of how and where these techniques should be used. The common view of having to wear some form of head mounted display to experience a virtual environment is a somewhat limited view because it is possible to experience a virtual environment by a number of other approaches - each having certain advantages over the other options. It is not possible to prescribe a particular option for all applications. Instead, it is necessary to carry out some form of human factors based analysis to determine user requirements. This is becoming increasingly important when we consider virtual systems because there are many more variables to deal with.

Detailed human factors analysis will reveal other important attributes of the design process such as communication between members of the design team, the customer, managers and designers. Collaborative working may be necessary when different branches of organisations are located around the globe.

3.4 Organisational aspects

Organisations who use computer based design systems will tend to see human factors in a different perspective. Here, factors such as productivity, reduced life cycle and cost of incorporating human factors into their process will be high on the agenda. However, to deal with all key issues, human factor considerations must be involved at every stage of the life-cycle from design and prototyping through to the utilisation of the product by the intended user. At each stage the human factors issues will be different but they will each have a major contribution to the usability and acceptance of the product. Consequently, the importance of human factors issues should not be under-estimated. In some cases competing products may offer exactly the same functionality but greater attention to the underlying human factors issues will make one product far superior to another. Less tangible, but nevertheless extremely important human factors aspects extends to product design where factors such as appearance are important. Unfortunately, it is very difficult to define in quantifiable terms the cost of ignoring human factors in a product. Overall, engineering design is a complex subject which frequently involves delicate compromises between ideal design, practicality, functionality, and market forces. The design tools used in the engineering-manufacturing process should be sensitive to these influences. The key to the validity of human factors lies in knowing how to measure the benefit.

4. HUMAN FACTORS IN THE DESIGN PROCESS

Human factors can play a crucial role in the design process by ensuring the resulting product is fit for purpose. One of the reasons why rapid prototyping is important is that it gives an opportunity for the design to be optimised by eliminating inconsistencies in the overall design.

5. CREATIVITY AS PART OF THE LIFE CYCLE

A key stage in the design of a product is the creative process where early ideas are converted into practical forms. The creative thought process is not fully understood except we know it to be a human characteristic stemming from man's earliest days on earth. The creative process is really a human factor. How many times have we heard the saying -"It was just a sketch on the back of an envelope"? Strangely, many manufacturers do not admit to this extremely important stage in the design process. At the first stage of a design being just a sketch there are almost an infinite number of choices in the realisation of the design but as the idea moves through the design process, options are honed down to a smaller set and eventually to one that suits the requirement. The 'sketch on the back of an envelope' is a very fast way of getting ideas communicated to others without necessarily having to worry about precise dimensions and structure. Ideas capture is very much tied into the creative process and does not easily fit the strict data entry protocols of many current CAD systems. It is interesting when one considers the creation process behind the UK's most successful vacuum cleaner - the Dyson Dual Cyclone. James Dyson made use of simple tools such as cardboard, scissors and cellotape to convert his original creative sketches into physical realisations of his designs. This illustrates that the creative process is not just conceptual, but that there are merits in being able to quickly prototype parts of the design in order to better understand the design in 'user terms'. Even though CAD systems make it easy to make changes there is a tendency to stay with a particular design because of the tedium when making changes. During the creative stage this is the very last thing the designer needs. There is a strong reluctance to scrap the design and start again. Figure 2 shows a product life cycle that recognises the creative process not only as a discrete stage early on in the overall process but as a continuous activity throughout the product life-cycle.

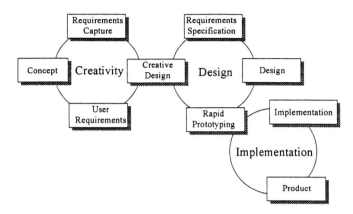

Figure 2: The Creative process as part of a product life cycle

Sadly, most CAD tools do not readily facilitate the early creative process because the interface and available functionality are too rigid. Even new generation tools offering digital prototyping are still too rigid and rely on data originating from conventional CAD systems. While CAD systems excel in their ability to represent a precise and accurate model they can be a barrier to the creative process rather than an aid. Unfortunately, systems offering a sketching capability lack the underlying functionality to make working prototypes.

Advances and acceptance in Virtual Reality (VR) concepts is beginning to lead to the availability of virtual prototyping systems. Division, a UK company have recently announced their latest VR software which supports building and interaction with Universal Virtual Products.

6. FUTURE REQUIREMENTS

An interesting trend is beginning to emerge around the use of networked high performance workstations. The most significant feature of these systems is the use of a network to share design data. Clearly, this opens up the possibility of a global network of distributed design teams. Future products could be designed and manufactured across the globe The global network looks set to become increasingly important and at least two companies have already taken steps to provide design tools that can publish virtual products over the Internet.

The provision of 3D fly-throughs for visualisation and exploration are becoming common place. In these systems the user can dynamically alter the viewing position to observe the design or see a complex animation taking place. However, several requirements are certain to remain the same in any future manufacturing company. They are based on the need to reduce the time to market and the man-power costs, the need to produce early product realisations and the need to offer flexible designs that can be rapidly tailored to suit the end user's needs. The time is coming when most manufacturing organisations will place increased reliance on virtual prototyping.

7. UNIVERSAL VIRTUAL PRODUCTS

A key step in any design process is to recognise the different groups of people who have an influence on the final product. Each user group will view the product from different perspectives depending on their role which could be any of the following: Concept Design, Detailed Design, Design for Manufacture, Sales/Marketing, Training and Maintenance (Refer to figure 3). Ideally, these collaborative groups should be able to interact with common engineering data in a way that suits their discipline. Many companies are part of an organisation where there is a need to communicate in a global sense. The advent of the world wide web has shown what can be achieved in terms of easier global communication.

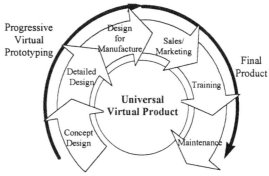

Figure 3:   **Universal virtual products**

A virtual product is to all intents and purposes a fully functional representation of a product albeit in a computer based representation. The virtual product will not only represent complex 3D assemblies but it will combine the behaviour and functionality of a real product -the only difference being that it will not exist as a physical representation. Importantly, it will be possible to revise and refine the virtual prototype until one is satisfied that it meets all customer requirements before the design is committed to the manufacturing process. The designer will feel confident knowing that many of the manufacturing and end user 'bugs' will have been eliminated prior to launch of the product.

8. CONCLUSIONS

It is very clear that virtual products will become an extremely important feature of future design systems. Producing a visual representation and a fully functioning virtual prototype is justified because it produces and achieves a better understanding of the product from conception of the idea through to the final development of the product.

9. REFERENCES

(1) **Preece J., et al,** (1994) *Human Computer Interaction,* Addison-Wesley
(2) **Shackel, B.** (1981) *The concept of usability.* Proceedings of IBM Software and Information Usability Symposium, Poughkeepsie, NY, 15-18 September. 1-30;

Virtual reality and operational design

C GRAY
Chief Executive, CADCENTRE Limited, Cambridge, UK

The term Process Plant encompasses any facility whose output is produced by a continuous or semi-continuous process, such as offshore oil & gas platforms, power stations, refineries, chemical plant and pharmaceutical factories. Process plants are usually large, complex and very expensive structures with many thousands of components. Interacting with all this data in a natural and efficient manner during the design cycle poses many interesting problems for the system architect. One tool which has proved very useful is Virtual Reality. Last April Cadcentre opened the world's first Virtual Reality Auditorium dedicated to process plant design. Mr Gray will review the use of Virtual Reality in designing tomorrow's world.

VIRTUAL REALITY AND OPERATIONAL DESIGN - J.C. GRAY

1. INTRODUCTION

Virtual Reality (V.R.) has fired the public imagination to a degree unmatched by any other recent new technology. The appeal of V.R. is not surprising : magical realism and the ability to enter imagined or fantastical worlds has always been popular, as witness the enduring success of "Gulliver's Travels" or "Alice in Wonderland". What the computer has brought about has been the possibility of actually passing through the looking glass - the flat screen of the computer terminal - and interacting directly with an imaginary world. Some commentators believe that V.R. will eventually have a greater influence on civilisation than computers have had up to now. Time will tell.

2. WHAT IS VIRTUAL REALITY?

The normal definition of Virtual Reality includes both a computer-generated three-dimensional world and the ability to interact directly with that world. This is quite a narrow definition because V.R. usually includes flight and vehicle simulators as well as arcade games and home entertainment. I would therefore include "reacting to" as well as "interacting with" in my definition of V.R. What is common to both definitions is the need to understand the artificial world before either reacting to it or trying to change it. Virtual Reality is, above all, about "Being There".

"Being There" is often achieved by means of a headset which cuts out the real world and projects the imaginary world onto screens in front of the viewer's eyes. Interacting with this space is easy if the controls are outside the imaginary world, as is the case for simulators or arcade games. It is much more difficult if the viewer's hands also appear in the V.R. space ("Virtual Dexterity") because then special gloves and/or body sensors must be used. It would be even more difficult to put more than one person into the V.R. space and to be able to simulate their interactions. For this reason flight simulators for commercial aircraft do not use headsets : they use a real flight deck with "virtual windows". Only the outside world is imaginary : the interaction between the flight crew with each other and with the controls must be real.

3. VIRTUAL REALITY AND PROCESS PLANT DESIGN

Cadcentre supplies engineering companies with software systems for the design of process plants. Process manufacturing can loosely be defined as "things made on a continuous basis" or "things which flow through pipes". (A more formal definition might be "a manufacturing process for products whose functioning does not depend on their external shape".) Examples of process manufacturing include : power stations; offshore oil & gas platforms; refineries; plant for making bulk chemicals; pharmaceutical factories; breweries; food & drink processing; waste reprocessors and water purification systems.

The total process industry sector is actually larger in global economic terms than classical discrete manufacturing industry. Civilisation as we know it depends on the process industries. Increasing their efficiency is vital to our future prosperity, not just in terms of personal welfare and consumption but also in terms of environmental management and improvement.

Process plants are usually very large structures consisting of many thousands of components. Offshore platforms in particular are very large and complex. The "Troll" platform for the Norwegian North Sea was the largest structure ever moved by man when it was floated out.

Some statistics for the "Mars" platform for the Gulf of Mexico give an idea of the sheer scale of a typical offshore platform. "Mars" is a so-called Tension Leg Platform : the entire structure floats on the sea, held in place by a network of steel pipes under tension which are themselves anchored to the sea-bed. The 5 topside modules (Power, Drilling, Process, Extraction and Accommodation for 130 people) each weighs about 2,000 tons. The modules are interconnected by over 400 stairways or walkways. The topside modules and flotation system together consist of nearly 50,000 components. The total project will cost £1 billion to complete.

These figures are impressive; but even a 15,000 ton platform such as "Mars" is dwarfed by the giant 25,000 ton "Ekofisk" platform in the Norwegian North Sea.

Coordinating the design of a large process plant used to be a formidably difficult task. In the past the design would be built up by teams of draughtsmen producing hundreds of Arrangement Drawings. To make sure that all the pipes, pumps, reaction vessels, structural steel etc. fitted together correctly and did not interfere with each other (for instance that a pipe did not run straight through a beam) the design team would then commission a scale model of the plant. These scale models could be huge - sometimes as large as the interior of a sports hall - could cost several million pounds and take many months to build, by which time they were usually out of date! Nevertheless, they were worth the trouble and expense because it is far preferable to correct errors in the design office, than to find out in the field that a design cannot be built and then to have to correct the mistake in some inhospitable or remote location.

About 15 years ago Cadcentre began developing a system which could replace the physical scale model of the process plant by a computer-generated digital model. Today this would be called "Virtual Prototyping" - except that in process plant design there are no prototypes because there are no production runs : each plant is unique. In process plant design "what you build is what you get". Hence the importance of removing as many errors as possible during the design process.

The computer-generated digital model has many advantages over the real-life scale model. It can be altered quickly as the design evolves; it can produce component lists which can be fed into procurement and costing systems; it can be used for calculations such as Centre of Gravity or steelwork stress analysis.

In its early days it was quite difficult to visualise a large digital model (more accurately, a digital design data base) because of the limitations of the graphics terminals available at that time. The breakthrough came at the beginning of the current decade when affordable, colour shaded 3D workstations appeared. The new workstations gave the project team the ability to see the evolving design, in colour, as it would appear as a constructed 3D process plant. The data base had become alive : what had previously been an incomprehensible jumble of wire-line images suddenly looked like the real world. It was also possible to program "walk-throughs" which would move the observer's viewing position through the model of the plant. Given the very large number of objects in the data base (each component of the plant was itself a small 3D model) the movement was rather jerky - several seconds per view, not the 30 views per second needed for a "real-time" effect. ("Real-time" could be achieved, but only by making a video of the walk-through and playing it back.) Also the representation, although geometrically accurate, was rather crude : simple colours, no textures, no backdrops such as sky or sea or ground. But compared with what had been available previously this was a revolution!

As is normal in the world of computing the rate of progress thereafter was rapid. In just a few years the price of the viewing station has fallen by more than a factor of 10, while the available functionality is vastly richer. With today's hardware it is almost impossible to distinguish a screen shot of a virtual plant from a photograph of the real thing. 30 views per second is also quite possible today.

Having found a way to view the project data base, it did not take long before the design process itself was operating in the new mode. As the definition of Virtual Reality includes the ability to interact with the imaginary world, we can safely say that as far as process plant was concerned, Virtual Reality Engineering arrived in 1992.

The scale model was now finally obsolete. But something had been lost : the ability for not only the design team but also the future operators of the plant to walk around the model, review its design and comment on its operability. It took a further 2 years before we were able to move from Virtual Engineering to Virtual Operations.

4. VIRTUAL OPERATIONS
One of our customers recently had a Fast Track project with a very demanding deadline. He urgently needed to speed up the operational review of a new offshore platform. This included checking the order in which the modules would be taken out to sea and lifted into place; the temporary structures which would be needed; and the clearances between these and the lifting barges. Another issue which concerned the design team was the safety of the helicopter deck which had an unusual location.

These operational reviews would involve people who were not familiar with detailed engineering drawings, nor with models viewed through a computer screen. They included the officers of the lifting barges : helicopter pilots : and government officials from the relevant safety organisations. The customer believed that some form of V.R. presentation would be ideal for getting the engineering issues across to a basically lay audience. The first suggestion was to give them all V.R. headsets. But this would have meant linking up 20 headsets so that each looked at the same object at the same time, which was technically very difficult to achieve in the short time available.

The solution was to do away with headsets altogether and use the IMAX principle instead. In an IMAX cinema the screen is large and curved. If the edge of the screen subtends a big enough angle to the viewer (about 60° on each side) then the viewer will suddenly "float into" the picture. Virtual Reality has been achieved, and not just for one person, but for the whole team. No cumbersome headsets or gloves are involved : the team can move about, make notes and interact with each other in a totally normal manner. In this instance the purpose of "Being There" was to facilitate the review team's understanding of the operational problems. Good interaction between the team members was more important than their ability to interact with the data (although the captain of a lifting barge became so carried away that he damaged the screen trying to walk through a virtual door!).

The results of this project exceeded our expectations. Initially the idea of using V.R. for an operational review met with considerable scepticism from the very people it was most intended to help. But scepticism gave way to enthusiasm as soon as the first major operational problems began to be identified, some of which could have seriously delayed the project. As a bonus several smaller issues, which would not normally have come to light were also noticed and rectified. The customer subsequently stated that without the use of V.R. it would not have been possible to meet the very tight project deadline, and that using V.R. had saved at least £3 million on a £70 million project.

C527/0

5. THE CADCENTRE V.R. AUDITORIUM

The project described above ran on the most powerful graphics computer available at the time, a Silicon Graphics Onyx Reality Engine 2 with three graphics subsystems each driving three high contrast projectors. The screen was 8 metres wide, and subtended 150° to the focal point. The equipment was located at the Silicon Graphics Reality Centre at Theale, near Reading.

Following the success of the project we decided to build our own Virtual Reality auditorium in Cambridge. As we wanted a system that could be installed in a typical office environment without extensive building works we opted for a smaller 11 square metre screen subtending 120° to the focal point. The screen is self-supporting, so the whole system can be dismantled and re-erected in a few hours. Our graphics system is similar to that used at Theale. Our 8-processor Onyx Reality Engine 2 can generate 100 million pixels per second, equivalent to 6 million fully textured triangles per second. Each screen shot requires 3.9 million pixels.

We inaugurated our Virtual Reality Auditorium, which we call our Visual Engineering Centre, in April 1996. It is the first V.R. facility in the world to be devoted to the process industry.

Having perfected the process of viewing and moving around in virtual plant space, we intend to continue our studies of "Virtual Operations". This will involve simulating the actual operation of the plant, with particular reference to hazard analysis; the validation of escape routes; the simulation of smoke, fire and explosions; and other safety related issues.

We believe that we made a significant breakthrough in the application of V.R. by using wide-screen projection in an auditorium rather than using headsets. One reason was that we were able to make our virtual world extremely realistic by using proper textures on the surfaces and a very high level of detail definition. The more realistically the virtual world is presented, the greater the sense of "Being There". And the more heterogeneous the group, the more important this becomes. Group Virtual Reality can greatly enhance the quality of their decision making by immersing them all in the same virtual environment. This concentrates their minds on the same problem at the same time. "Being There Together" is the common focus.

6. THE FUTURE

We can be quite certain of two things. First, that Virtual Reality, even with today's technology, can offer real benefits in worlds other than entertainment. Indeed, some analysts predict that by AD 2005 entertainment will account for only 40% of all VR applications instead of over 80% today. Second, that the rate of improvement of the hardware technology will continue to increase exponentially. Back in 1965 Gordon Moore, co-founder of Intel, observed that the number of transistors on a chip was doubling every 18 months without a limit in sight, which could give us one billion transistors on a chip by AD 2010! Today computer power per dollar is increasing by a factor of 10 about every 3 years. At Cadcentre we have seen the graphics performance of our own V.R. system increase by a factor of five in just 12 months!

A greater restriction on the growth of V.R. will be the rate of improvement of the projection system. More progress in plasma, thin-film and panel display technology is needed before the promise of Virtual Reality can be fully exploited. Even more work is needed on gloves, sensors and other so-called "haptic" devices before "Virtual Dexterity" becomes a tool suitable for everyday use.

As I noted in the introduction to this paper, some experts believe that Virtual Reality will have a greater influence on civilisation than computers have had up to now. Others fear that Virtual Reality, like Artificial Intelligence before it, will fall victim to its own hype and become discredited. The truth is probably more mundane. Technical progress is only continuous in hindsight. It usually proceeds in bursts, with significant pauses while some component waits for the breakthrough that allows the pace of the advance to resume. It will be the same for Virtual Reality. The only thing that is certain is that the techniques and equipment which we use today will hardly be fit for a museum by AD 2010.

C527/0

signing future adaptable transport for the needs of their
ers

WKES BSc, MErgonSoc
r Consultant, MIRA, Nuneaton, UK

SYNOPSIS

During the 20th century road transport has become the most popular means of providing personal mobility throughout the world. Modern vehicle design considers the needs and abilities and of the driver and passengers in great detail to produce cost-effective vehicles with high standards of ease of use, comfort and safety. The consideration of human needs in design is driven by the knowledge and techniques developed by the applied science of ergonomics. This paper surveys the development of vehicle design for the needs and demands of its human users by reviewing the development and the future direction of vehicle ergonomics.

1. INTRODUCTION

In a historical sense, self-powered road vehicles are a relatively new phenomenon. Although important antecedents existed in the late 18th and early 19th century, it was not until the work of pioneers like Karl Benz and Gottlieb Daimler in Germany that a recognisable "car" (1887) and "motorcycle" (1885) were constructed. While the mechanical ancestory from these vehicles to those of the present day is clear, the process of conceiving, developing and manufacturing vehicles today has changed radically. Over the last 100 years many new performance requirements and operational constraints have been progressively added that significantly influence this design process. Some of these are legal requirements for vehicle design that relate to vehicle safety and environmental performance. However additional targets defined by the needs and expectations of the eventual customers are also important. The development of vehicles that offer a cost-effective competitive blend of features, performance, comfort, economy and safety is therefore a common goal for the vehicle designers and manufacturers.

Therefore the manner in which these dissimilar targets are represented and the way in which decisions are made on any necessary trade-offs throughout the development process are crucial to the eventual outcome. To understand how the needs of the human users are considered during vehicle design we have to consider both the development of the world market for transport and how targets are set that relate to human needs in developing future vehicles.

2.   THE VEHICLE AGE

Although road vehicles exist in many forms the dominating influence is undoubtedly the passenger car. It is estimated that at the end of 1995 there were 24.96 million cars registered in the UK, 161.35 million in the European Community and 492.73 million across the world (1). However the distribution of road vehicles across the globe is by no means uniform. Approximately 79% of the total world parc of cars is being used in the combined territories of the EU, North America and Japan, which only has 14.56% of the world population. This statistic indicates nearly 2.2 humans per vehicle in these most highly developed areas. Predictions of future growth of the vehicle parc across the globe indicate that those areas currently with relatively low traffic density, i.e. continental Africa, South America and Asia, are likely to markedly increase while those that are currently heavily populated areas will also increase. The result is that the future vehicle parc will continue to develop and the nature of the global driving population will change as diversification occurs. However increasing vehicle populations and usage has other known effects on society. Congestion, pollution and accidents are all present day by-products of road transport growth. Identification of the growing nature of some of these difficulties has motivated both governments and manufacturers to identify specific priorities for future vehicle and traffic performance that require technology and policy solutions. As many of these proposed solutions will be directed towards the driver's use and behaviour with his vehicle it is important for us to understand how drivers will react to them before they become embodied in future vehicle design.

3. THE DEVELOPMENT OF ERGONOMICS

The science of Ergonomics seeks to enhance the effectiveness of the use of tools and equipment while maintaining an emphasis on achieving desirable human values (e.g. health, safety and satisfaction). It achieves this by applying information about human characteristics and behaviour to the design of items that people use and the methods by which they are employed. It is therefore an applied area of science that combines the knowledge and techniques developed in the fields of psychology, biology, mathematics, physics and engineering. This field of study is also commonly termed "Human Factors Engineering" and in the context of vehicle design it relates to all aspects of the vehicle that influences its interface with the human occupants. The rise of this field of study has followed the development of industrial society. McCormick and Sanders (2) have described some of the historical aspects of this development. Early approaches were within a military context. The necessity to enhance the "performance" of humans during the First World War forced consideration of the need for adequate selection and training procedures for *"fitting the man to the job"*.
The related pressures of military conflict in the Second World War brought another step change to this emerging field of study.

As new innovative technologies and systems emerged it became apparent that some were ultimately limited in their performance by the abilities of their human operators to control them effectively. This realisation brought with it a change of emphasis that is defined in the redefined title *"fitting the job to the man"*. Consideration moved from the physical characteristics of equipment and control, often characterised as the *"knobs and dials"* phase of ergonomics, to eventually encompass other wider issues such as safety of operation and allocation of tasks between man and machine. The relevance and importance of a human centred approach to industrial design has now become widely accepted in consumer product design and ergonomic assessments for usability and safety are commonly embodied in product requirements.

4.   VEHICLE DESIGN AND ERGONOMICS

All of the stages of ergonomics development described above are reflected in the way in which the approach has been applied in the vehicle design process.

4.1   Gradual development

As the earliest road vehicles emerged from the horse drawn cart, it is unsurprising that the driver of the turn of the century car conducted his vehicle from an open cart-like bench seat with very simple mechanical controls. Tiller steering and multiple levers together with some rudimentary gauges provided what control and information was thought necessary. These early vehicles were bespoke craftsmen built assemblies and evolved from their designers experience of operating the vehicle. Incremental refinements such windscreens, enclosed cabins, gear levers, more effective pedal operated braking systems and steering wheels eventually allowed the driver some greater comfort and control over his vehicle. As these features, and others, were added, the conventions of the modern vehicles driving compartment began to emerge. Mass production techniques and the dictates of the assembly line in the 1920's began to necessarily influence the "standardisation" of these features. However by the 1930's vehicle designs were still being produced with only a rudimentary definition of the space and comfort requirements of the eventual customer. As a result of the experience of the second World War and military equipment development the automotive industry adopted new philosophies of representing human user characteristics as *geometric* models in the design process. By the 1960's acknowledgement of the growing numbers of road accidents brought a particular focus on how vehicle design could be improved to reduce occupant injury after such an event. This stimulated a continuing line of research into human *bio-mechanics* and the way in which injuries are caused and how the human body could be protected in a vehicle environment. Moving to the present day, we now see a greater demand to understand how the *behavioural* attributes of the driver population can be accounted for in future vehicle design. As active systems to inform and assist the driver emerge then it is important to understand how drivers will accept and adapt to them.

4.2   Characterising the population

Clearly an essential piece of information for the vehicle designer is some appropriate definition of the physical characteristics (e.g. size, shape, weight etc.) of the population for which he is designing the vehicle.

The availability of such basic anthropometric data is not a trivial issue as many early sources of data related to military populations only which were not necessarily representative of a civilian population with different demographic characteristics. Individual studies such as that carried out by Haslegrave (3) which surveyed the body size of 2000 UK drivers and front seat passengers in the late 1970's allowed the definition of a static anthropometric data set from which human population target sizes could be defined for vehicle designers.

4.2   Packaging in the passenger compartment

The availability of such data on human variability facilitates the first important step in the design process, i.e. to define a volume within the overall vehicle structure that is reserved for the occupants. This occupant package is then detailed as the placement of components (e.g. seats, controls, ventilation outlets, body structure etc.) is gradually defined and their mechanical interface with the driver and passenger assessed. It is essentially an iterative process. Rebiffe (4) identifies the detailed procedures by which this geometrical evaluation occurred in traditional two-dimensional layout work in vehicle design. He identifies the way in which this anthropometric data on limb length, stature, reach etc. can be turned into an engineering design analysis where both accommodation within the package and comfort of the seated occupant is considered. However a later discussion on occupant packaging by Roe (5) describes some of the advantages and practical difficulties in converting such human data into useful assessment techniques to evaluate vehicle designs. He also describes the development of some physical tools to aid the designer. This includes the definition of both the two-dimensional and three-dimensional manikins that were developed in the 1960's and are still used within the automotive industry today. These surrogate representations of the human body assist assessment of vehicle packaging and attitude comfort for the seated "occupant" on layout drawings and prototypes. They also form the basis for many other human related aspects of vehicle design and regulation. The three-dimensional manikin and its installation procedure allows definition of a Seating Reference Point (SgRP) within the vehicle around which many performance aspects of the design are assessed. Modified versions of these manikins have been used in applications as diverse as assessment of pressure distributions across the seat of the occupant and driver, visibility of both controls and displays and the external environment via "eye-points" derived from this Seating Reference Point (6).

Obviously Computer Aided Design (CAD) approaches are now normal in the industry and these practical assessment tools are now supplemented with CAD equivalents. Many man-modelling systems have been developed to allow manipulation of human body shapes within a computer environment to assess packaging issues. Porter (7) summarises the use and application of such systems in automotive design. As such models become more refined with more human population data, it is likely that they will become a key feature in future vehicle development. Virtual reality techniques currently under development will undoubtedly also play a part in evaluating and refining the physical design of the future vehicle.

4.3   Comfort, Convenience and Safety

After considering the overall packaging of the vehicle passenger compartment there are many other vehicle features that in detail design contribute to how comfortable the driver perceives that environment to be and how successfully the vehicle can be operated. Essentially the same principles can be said to be applied.

Firstly the performance target for the new design has to be set. As vehicle design is essentially a market driven process, this will commonly include user related performance targets based upon competitor vehicles performance. This can consist of both objective measures and subjective ratings drawn from panels of representative drivers. These so called benchmarking exercises frequently offer a first step to setting user related targets to comfort and convenience, issues that can then be supported with more detailed ergonomics studies of the effectiveness of detailed aspects of vehicle design. The internal environmental performance of the vehicle is also an important consideration to be addressed. Noise, vibration, heating and ventilation issues are also ones where data from ergonomic assessments of driver performance and preference studies can offer meaningful initial design targets that are subsequently supported by subjective rating studies by users.

It has already been noted that design for effective safety has been a particular emphasis in vehicle design, both primary safety (i.e. avoiding the accident) and secondary safety (i.e. surviving the accident) are dual complimentary approaches. Up to the current day refinement of secondary safety systems in particular (e.g. seat belts, deformable steering wheels and air bags) has required considerable knowledge of human bio-mechanics and both theoretical modelling and data collection with anthropometric test devices, or crash test dummies as they are more commonly termed.

5. FUTURE VEHICLE TECHNOLOGY

Throughout its history there are many examples of how automotive design has absorbed and assimilated new systems and components to address emerging needs and increase some aspect of vehicle performance. Current perceptions of society needs for road transport have been reviewed (8). This analysis suggests that the key areas related to the human operator of the vehicle are and increased customer expectations for more comfortable, convenient and safe cars, avoidance of traffic congestion and improved environmental performance.

Research programmes like PROMETHEUS in Europe and similar activities elsewhere have identified systems that could make significant contributions to addressing these real world needs. However these will only be proven if they can be developed for a mass automotive market and they can be shown to be effective in operation. Many of these concepts involve the development of advanced electronic control and information systems for vehicles and drivers with inter-vehicle and roadside communications. Parkes and Franzen (9) report on the development and implications of such technology. This review indicates that future vehicle concepts under development could lead to on-board systems that can assist driver decision making by offering drivers *information* and *assistance* services.

Improved sources of *information* via vehicle geographical databases that can be updated by real-time broadcasts are conceived and are nearing the market. These systems can allow direct route guidance advice to be available to the driver. However additional complex information availability may not necessarily be beneficial at all times to all drivers. Consideration of format, availability, timeliness and accuracy of such information is key to its effective and safe use by drivers. This will equally apply to other information services potentially relayed into the vehicle by advanced communication channels where an important feature must be a prevention of driver distraction.

Should the driver control the speed and delivery of this information and response medium, or should the system be so designed to manage the dialogue between itself and the driver? When we consider proposed *assistance* systems to the driver for longitudinal or headway control (Autonomous Cruise Control), and lateral control (lane departure and overtaking) we see the potential for different layers of system advice and/or intervention. This can range from systems that warn the driver only through ones that provide advice and some intervention to ones that finally take over control from the driver in hazardous situations. While potentially of benefit such concepts raise fundamental questions about the role and responsibilities of the driver.

All of these technological possibilities have to be seen in the context of the changing demographics of the global driving populations. Predictions suggest that greater proportions of both female and older drivers will be present in the future driving populations. The US National Highway Traffic Safety Administration has estimated that by the year 2000 80% of people aged over 70 will have a driving licence. The abilities and needs of this changing market needs to be considered in our future decision making towards any new technology.

6.   CONCLUSIONS

Refinement in vehicle design to meet the needs and expectations of their users has been based upon appropriate definitions of human variability being considered throughout vehicle development. After geometrical, spatial, bio-mechanical consideration to human requirements have been given we are now considering how the behavioural aspects of human users can be assessed within new vehicle development. The emerging technologies to address real world problems may offer a fundamentally different driving environment for the future. Continued evaluation of the characteristics of future drivers and their interface to the vehicle environment is essential to ensure that claimed benefits are maximised.

References

1) "Motor Industry of Great Britain 1996 - World Automotive Statistics"
Society of Motor Manufacturers and Traders Ltd, London 1996
2) "Human Factors in Engineering Design", McCormick, E J & Sanders M S,
McGraw Hill, 1983
3) "Anthropometric Profile of the British Car Driver", Haslegrave, C M
Ergonomics, v23, May 1980, p437 - Taylor & Francis, London
4) "An ergonomic study of the arrangement of the driving position in motor cars", Rebiffe, R
in "Ergonomics and Safety in Motor Car Design", Institute of Mechanical Engineers, 1966
5) "Occupant Packaging", Roe, R W, in " Automotive Ergonomics", ed Peacock, B and
Karwowski, W - Taylor & Francis, 1993
6) "The legal determination of the drivers field of view", Fowkes, M, in "Vision in Vehicles",
North Holland, 1986
7) "Computer-aided ergonomics design of automobiles", Porter, J M et al
in " Automotive Ergonomics", ed Peacock, B and Karwowski, W - Taylor & Francis, 1993
8) "CAR 2000", Motor Industry Research Association, Nuneaton, UK, 1993
9) "Driving Future Vehicles", ed Parkes, A M and Franzen, S - Taylor & Francis, 1993

ultimedia information systems in manufacturing

•fessor W HALL BSc, MSc, PhD, CEng, FBCS
fessor of Computer Science, Department of Electronics and Computer Science, University of Southampton, UK
R CROWDER BSc, PhD, MIEE
ıior Lecturer, Department of Electrical Engineering, University of Southampton, UK

The management of information is one of the greatest problems facing industry today. This includes both the integration of information from various sources, and making relevant information available to managers and engineers when and where it is needed. This paper will look at ways in which multimedia information systems can be used to improve manufacturing efficiency, including the use of novel interfaces to complex information systems and the use of software agents to improve access to such systems. The paper uses an example of work being undertaken at Southampton to illustrate the points being made.

1. INTRODUCTION

It is well know that within manufacturing industry, a substantial amount of unnecessary time and effort are expended by engineers and support staff looking for, waiting for and copying engineering data. In addition there is a lack of visibility across organisations of design data and progress information. These difficulties can be seen to be significant contributor to higher costs, delivery problems and relatively poor productivity. In response to these problems a number of solutions have been proposed including the on-line electronic register of engineering documentation [1], all electronic manuals and the use of open hypermedia [2]. However it is inconceivable that with the current availability of network technology, the introduction of an electronic document management system, EDMS, can be seen as a stand alone activity. It must be viewed as part of a total integration strategy between a company, and its customers and suppliers. This leads to the concept of the virtual factory, where a number of small manufacturing centres, each focusing on their core competencies, combine to manufacture a product [3].

This structure will enable the factories to operate as one regardless of geographic location, giving an increase in flexibly and reduction of costs. Developments in network technology will make it possible for companies with dissimilar computer systems to exchange information for example about design, inventory levels, the manufacturing process and delivery schedules. In principle it will allow all manufacturing partners to have equal access to information. However for many companies the development of systems and business processes that allow total electronic collaboration remains elusive. Even highly sophisticated companies have found, and continue to find, that the task of creating seamless electronic networks of lean computer integrated manufacturing operations to be frustrating and difficult, [Upton 96]. Managers of most companies are struggling to make their information systems more flexible, and are perplexed about why so much paper is still being shuffled around both the design office and the shop floor. Two critical elements make this type of networking possible, a function termed the information broker and the open standards based on the protocols established for the Internet and the World Wide Web (WWW). These open standards make it relatively easy for members of a community to share information regardless of differences in their individual IT systems and geographical location.

Partly as a result of these developments, there is a growing demand for the provision of globally distributed engineering information. Many organisations have been operating on a global basis for many years, and are increasingly supported by intranets designed to their specific requirements. Access to knowledge at different locations stimulates co-operation between industries, and forms the basis for the virtual enterprise. The virtual enterprise implies teleco-operation between organisations via networks and services. The partners in the virtual enterprises are working together in a commonly agreed time frame, which may be very short, with a common goal.

Current engineering practice, particularly concerning the early design stage of the product, requires redefining to make full use of the rich diversity of available and emerging information technology. Such technology can address a wide range of activities including the capture of the requirement definitions, product modelling, and the product's complete documentation (drawings, manufacturing procedures, third party information etc.). Currently the basic components required for this strategic communication, such as hypermedia and network technology, are available to exchange information electronically. However the common services and methods used to exchange information over wide area networks, and to facilitate a range of supporting collaborative work are currently lacking. The information generated during the design process will form the core of the information required during the life cycle of a product. Engineering practice is evolving as companies are forced to increase their overall effectiveness, flexibility and response times to customers. To overcome these challenges the company must maximise the reuse of existing internal and external information, while forming virtual enterprises with firms which focus on other core competencies.

A successful engineering information management strategy must be able to integrate information, irrespective of format, from a large number of sources, and allow personnel easy, and controlled access to the information. In addition the integrity of the information must be maintained by the implementation of suitable change control, and associated security measures. Finally if documentation is to be changed, the required process must be error free, and procedures used must be capable of auditing by the use of recognised quality procedures. This paper presents an approach to the optimisation of multimedia information resource management by employing open hypermedia technology. ~

2. MULTIMEDIA INFORMATION WITHIN MANUFACTURING
To the manufacturing engineer, integrated manufacturing means the integration of CAD and CAM, and to the shop floor it means a communication and integration exercise. To

management true integration is concerned with the global information systems in which all parts of the organisation base their operational activities on a common and current data in a shared database. Shared data has been termed the integration glue that binds the system together and allows it to operate as a co-ordinated whole. Integration at the enterprise level rather than the technology level, confined to the shop floor or drawing office, is the objective of the information systems view of integration.

Database and information systems technology therefore have a crucial role to play in modern manufacturing management. Rigorous and consistent structuring of data that describes products and the flow of associated product data from design to final shipment can potentially allow anyone within the organisation to access the particular items of data they require for their decision making process currently most industrial functions of a manufacturing organisation maintain their own private database. Rather than dispense with these separate databases in favour of a single common database, it is likely that individual databases will be retained but with a common interface.

Access to stored common data will not by itself be sufficient, however. Management will need an increasing amount of intelligent support from the computer system in the retrieval and interpretation of the relevant data, and in guiding the appropriate course of action. In the fields of manufacturing, intelligence is generally given to individual shop floor devices. The next logical stage is to give distributed computer based intelligence at a higher stage of the enterprise, by capturing and structuring the knowledge employed by the engineers in making technical and operational decisions at a higher level of abstraction and incorporating this into the knowledge base.

3. APPLYING OPEN HYPERMEDIA TECHNIQUES TO MULTIMEDIA INFORMATION MANAGEMENT

At Southampton we have been using open hypermedia techniques to support the development of an information management environment as described above. Most people today are familiar with the concepts of hypertext and hypermedia through the increasing popularity of the World Wide Web. However, common usage of the Web embeds the information about the hypertext links between documents in the documents themselves. Whilst increasingly documents in large-scale Web sites are being managed using database technology, linking information is still only available as embedded references in the documents, and can only be used if the document is in the Web html (hypertext mark-up language) format. This becomes unmanageable in large-scale applications and means it is very difficult to integrate legacy data into the information system unless it is all converted into an html compatible format.

Open hypermedia systems typically store links as entities in their own right, separately from the document data, so that they can be managed and processed in the same way as we manage and process documents. They also typically provide a link service that can be applied to many different document data formats including for example legacy data. At Southampton, we have developed the open hypermedia system Microcosm [4] which is based on a three layered model which is similar to the three-level architecture of the traditional database management system (DBMS). In its basic form, the architecture of Microcosm permits three levels of access to multimedia information, namely - browsing (hypermedia link service), unstructured (such as information retrieval) and structured (such as the hierarchical catalogue structure in the document management system).

Microcosm[5], [6] is a hypermedia system which allows users to browse and query large multimedia collections. However, Microcosm is fundamentally different to most hypermedia systems because of its ability to integrate information produced using a variety of third-party applications. It consists of a number of autonomous processes which communicate with each other by a message-passing system . No information about links is held in the document data files in the form of mark-up. Instead, all data files remain in the native format of the application that created them. All link information is held in link databases (linkbases), that hold details of the source anchor (if there is one), the destination anchor and any other attributes such as the link description. This model has the advantage that it is possible to have different sets of links for the same data, and also it is possible to make link anchors in documents that are held on read only media such as CD-ROM and videodisc.

Due to its open structure, Microcosm permits the integration of data generated by suitable third party applications, for example CAD packages, databases, and spreadsheets, into an integrated information environment. This feature enables users to access the full capabilities of these individual applications, whilst still having access to multimedia resources containing information, that describes or enhances the data being manipulated. An example of this is the integration of Microcosm and the commonly used drawing package, AutoCAD. This allows the user to select any component on the circuit diagram, and link directly from it to access details such as ordering information, maintenance information, or its actual physical location in the machine.

4. A CASE STUDY IN FACTORY INFORMATION PROVISION: PIRELLI ABERDARE

We have been using Microcosm for a number of years to develop multimedia information systems within the manufacturing industry. One of our main case studies has been undertaken with Pirelli Cables. The main product at of Pirelli Cables at Aberdare is standard three core building cable, a large proportion of which is sold in packs containing 100m lengths. To form the packs, a number of packaging lines are used. The mobility of labour at Aberdare and the fact that the packaging line is by its very nature, complex, made it an ideal candidate for use with a hypermedia maintenance resource.

The biggest cost in the implementation of a new hypermedia application is in the authoring. This requires significant effort in several areas, including the collection and formatting of the data, constructing and checking the linkbase, and subsequent maintenance of the application. Collection and formatting of the data is the biggest element and can vary depending the available source material. At one end of the spectrum the documentation to be used may only be available in a paper form and therefore can not be directly loaded into the application. At the other end of the scale all the documentation may be available in electronic form making the authoring a quick and easy task. If the source material is largely paper based, we have estimated that up to 80% of the time required to develop a new hypermedia application is taken in converting the information into a suitable format.

Once a data set has been selected and collected the information needs to be structured in such a way as to make navigation through the application easy and logical. Links are then overlaid on the data and highlighted, in the main, as buttons which are obvious graphical hints to the user where they can move to in the database. In the case of engineering drawings it is considered more efficient to define links to the locations of parts as generic links using the name of the part. This has the added advantage in that the part location can then be accessed wherever the part name is used without any extra authoring effort. Using AutoCAD for the circuit diagrams and engineering drawings allows Microcosm to recognise components as objects rather than as a collection of pixels bounded by a set of rectangular co-ordinates. This ensures that all the links will keep their meanings whenever changes are made without having to re-author all the links on a diagram. We are now moving towards using the DXF format for diagrams and drawings which permits all the advantages of AutoCAD for link creation and maintenance, whilst not requiring a copy of AutoCAD to be available when the application is running.

One of the main differences between traditional documentation and hypermedia is the ability to use temporal data such as sound and video. These have to be used carefully as the user can often become frustrated at having to sit through the same data again and again to get to a particular piece of information. We have also developed a 3-D model of the machine as an interface to the information system, which is very useful for novice users but more likely to be by-passed by the experts.. The application developed for Pirelli Cables is capable of providing information to support maintenance and other engineering activities in response to direct requests from users. It provides information for a number of activities including set-up, maintenance, diagnostics and training. Design requirements included the presentation of information in a clear and consistent manner, with an easy to use interface, capable of being operated within the factory environment, without compromising the safety of the user. In early trials a portable PC with a pen interface was used, which enabled the operator to access information whilst working on the line. We are now experimenting with the new generation of hands-free "wearable" computers which include speech interfaces. Whilst offering many advantages, the problem of noise in the factory environment cannot be underestimated when using speech interfaces. The equipment is also still too heavy for operators to wear for any length of time.

5. FROM INFORMATION FILTERING TO AGENTS

The Microcosm filter architecture allows for different processes to be applied to dynamic link creation, such as information retrieval and rule-based algorithms. This facility can be used for the automatic generation of links in large resources bases, together with the development of intelligent 'agents' to guide users around the resource base, for example during fault-finding within industrial plant. As the linkbases are implemented as filters in the Microcosm model, they can be inserted and deleted from the filter chain very easily. This means it is possible to create different hypertext views (for example those required for training or maintenance) of the same set of data, and in addition configure the current view dynamically during real time diagnostics.

A computer-assisted fault diagnosis procedure was built to complement the application described above [Heath, 95]. This can be used to support engineers when real faults occur on the machine and to simulate faults for training purposes. The rule-base is a simple set of facts, with some rules which provide access to those facts. A Prolog filter was used within

Microcosm to create the rule-base. When the fault finding process commences, the user interface sends a pre-defined clause which provides the gateway into the rule-base. The result of this is then presented to the user as a simple form. The user can select any of the possible answers to the query, at which point the Prolog clause associated with the answer is evaluated, resulting in another form being displayed with another set of questions or possibly list of potential faults. These faults act as a gateway into the Microcosm document and link set. When a list of possible faults is displayed, the user is able to follow links from each fault to related information, perhaps giving the location of sensors which need to be tested to determine if that fault is indeed the problem.

6. DISCUSSION

If this hypermedia approach is to be fully adopted to deliver operation and maintenance documentation, the logical extension, as we have discussed, is to take the delivery system on to the factory floor to supplement the diagnostic process, and not solely as a computer based manual. In this case, the operator will receive assistance from the hypermedia system during the investigation of a fault. While the use of a portable pen computer is acceptable, a number of problems need to be considered, in particular, the security of the computer with the information stored on its disk drives, and the inability to communicate with the factory wide communication system. Recent advances in communication technology have permitted the development of radio based networks, and remote user interfaces.

When an interface to a factory based local area network is available, a large number of significant possibilities arise. In particular the breakdown maintenance process can be significantly enhanced, both in time and effort, by the implementation of intelligent filters, using information both from the database and from the failed system obtained through the network. A direct result of this approach is that when a fault occurs, the resource base is entered at the appropriate point (for example giving the location of a component and its functionality), as identified by the intelligent filtering process. Once this is achieved the hypermedia information system will form an integral part of the computer integrated manufacturing system. This communication linking will allow predictive and preventative maintenance procedures to be optimised by using the same resource base but with a different linking strategy. The advantage of the integration of a multimedia package hypermedia package with a network can not be under-emphasised. It will additionally give the user the facility to access remote data bases, for example stores' records and project management activities, to ensure that problems are solved rapidly, and efficiently.

While the paper has identified problems with the conversion and collection of paper based information, we consider this not to be a significant problem in future applications. With the universal use of word processors and CAD systems, the generation of hypermedia documentation for maintenance, training and diagnostic applications will become a seamless additional exercise task during documentation preparation. The application we have developed, is currently being used at Aberdare, and is proving to be beneficial to the maintenance, and fault finding activities. Currently we are setting up additional trials to quantify the benefits to the company.

In this paper we have discussed the use of open hypermedia systems to develop a resource-based approach to the provision of electronic information systems for the support of maintenance operations in a large-scale industrial plant. The advantages of the open system approach are in the maintainability of the resource-base, the automatic generation of links and the consequent reduction in authoring effort, and the ability to integrate the resource base with other application packages and the general computing environment. The case study described is now undergo a full evaluation at the Pirelli site in Aberdare, but our initial trials have already clearly demonstrated the advantages of the system, and its widespread acceptance.

One additional question that needs to be raised is who will determine the move to multimedia information provision; the equipment supplier or the user community. It is our view that the initial pressure must come from the user community. As we have noted to be viable the hypermedia resource base must operate factory wide, and not centred on a single machine, hence the incorporation of this technology must form part of a company's overall information strategy.

6. ACKNOWLEDGEMENTS

The authors acknowledge the EPSRC for funding the work under grant GR/H43038, and Pirelli Cables Aberdare for permission to use their site as the case study

7. REFERENCES

[1] Hogbin G. and Thomas D. Investing in Information Technology, McGraw-Hill, 1994

[2] Malcolm K, Poltrock S and Schuler D. Industrial Strength Hypermedia: Requirements for a Large Engineering Enterprise. Proceedings of Hypertext'91. ACM Press, December 1991.

[3] Upton D and McAfee A. The Real Virtual Factory. Harvard Business Review, July-August 1996

[4] Davis, H. C., Hall, W., Heath, I., Hill, G.J. and Wilkins, R.J. Towards an Integrated Environment with Open Hypermedia Systems. Proceedings of the ACM Conference on Hypertext, ECHT'92, Milan, Italy, December 1992, pp 181-190. ACM Press 1992.

[5] Hall W. and Davis H. C.: Hypermedia Link Service and their Application to Multimedia Information Management. Information and Software Technology, Vol 36 No 4, 1994.

[6] Hall, W., Davis, H.C. and Hutchings, G.A. Rethinking Hypermedia: the Microcosm Approach. Kluwer Academic Press, 1996.

[7] Crowder, R., Hall, W., Heath, I. & Bernard, R. "The Application of Large-Scale Hypermedia Information Systems to Training". Innovations in Education and Training International ,Vol. 32, No.3 pp 245 - 255 (1995)

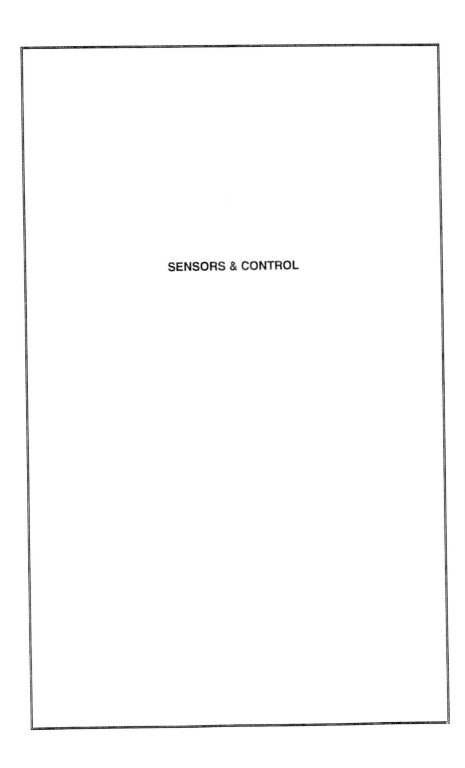

SENSORS & CONTROL

tomotive telematics

essor **J D TURNER** BSc, PhD, CEng, MIEE, MInstP
anical Engineering Department, Southampton University, UK

Synopsis.

This paper describes the current state of development of vehicle telematics, and reviews its likely future development and applications. The use of IT and electronic control systems ("vehicle telematics") to advise the driver and/or fully or partly control vehicle functions is described. Vehicle telematics can be subdivided into five main areas of application: vehicle control & driver assistance, route guidance/optimisation, environmental protection, fleet control, and road pricing. An overview of each area is presented, together with an outline of the likely future course of development in each area.

1. Introduction.

Since about 1950 the car has evolved from being a loosely-regulated consumer item, often seen as an indicator of personal wealth, to its current status as a near-universal and almost essential possession. Car, truck and bus users are subject to increasing levels of regulation, with the aims of curbing road use, controlling parking, environmental protection, and ensuring that vehicles are adequately maintained.

Traffic levels continue to rise, and in the UK vehicle numbers are predicted to double by early in the next century. The initial response of most national authorities to rising traffic levels is to build more roads, but this has become increasingly unacceptable. Throughout the world, the climate of opinion is turning against unfettered mobility, and the use of information technology (IT), together with electronic control systems to advise drivers and/or control vehicle functions is being seen as at least a partial solution to current and predicted traffic problems. The combination of IT and vehicle or highway-based electronic systems is known as telematics.

2. The reasons for introducing telematics to road transport:

- First, to control road use and to spread demand away from peak times. Some newly industrialised countries, such as Singapore, have brought in sophisticated measures to control road usage. Others (e.g. Thailand) will have to tackle city congestion urgently.

- For environmental protection, to help control global warming (CO_2), acid rain (NO_x), hydrocarbons and particulates (both of which are health hazards). In the United States the California Air Resources board (CARB) is applying pressure on vehicle manufacturers to develop "zero emission" vehicles, or at least virtual zero emissions, whereby vehicles contribute no more pollution to the local environment than nearby power stations .

- Two recent reports in the UK have highlighted the environmental impact and congestion problems due to road vehicles:

    - The House of Commons Select Committee on Transport reported on environmental pollution [1], and concluded that the introduction of lead-free petrol was not universally beneficial when the complete UK vehicle fleet was considered, and the effect of other fuel additives taken into account.

    - The Royal Commission on Environmental Pollution [2] called for a radical overhaul of Britain's transport policies, particularly by adjusting investment and the taxation regime to favour public transport.

- The construction of new roads is becoming increasingly unacceptable to the public, as demonstrated in the UK by the debates surrounding the construction of the M3 through Twyford Down, Newbury Bypass, and, most recently, Salisbury Bypass.

- There are early signs of research findings from around the world linking respiratory diseases with emissions, and especially particulates, which are currently unregulated.

The next decade is likely to see widespread public debate about the balance between personal mobility and public responsibility. With this debate will come many political and social challenges to long-held values and accepted rights. It is likely that, over the next decade or two, road vehicles and their use will become radically different, as a new balance is achieved between added sophistication, performance, telematics and driver aids, and reduced weight, emissions, construction and recyclability.

Many of the problems caused by road vehicles are capable of alleviation, if not total solution, by the application of transport telematics. Telematics is increasingly being seen as one of the keys to sustainable mobility [3]. Congestion is already worse in the UK (where on motorways we have more than 1200 vehicle kms travelled per km of road per hour) than in many other developed countries.

3. The development of automotive telematics.

The control of congestion and the efficient use of vehicles requires intelligent vehicle and highway systems which communicate with each other. It is going to become more and more difficult to consider the vehicle or highway systems in isolation: they will increasingly interact. The introduction of features such as navigation and route optimisation systems, together with the use of network-derived messages delivered to the vehicle, will increasingly affect the traffic flow.

It is interesting to note that, in many cases, the required technology already exists, or at least can be easily developed from systems currently in use. Most of the problems which have to be addressed before vehicle telematics can be widely introduced centre around legal issues (such as liability in the event of an accident), customer acceptability, or unwillingness on the part of vehicle manufacturers or highway authorities to make the necessary investment. Part of the problem is to do with the perception of risk: considerable effort will have to be put into persuading drivers that surrendering some of their authority over a vehicle is acceptable. For some reason society tolerates the fact that human drivers cause several hundred deaths a month in Europe - but there would be an enormous outcry if it were found that telematic systems were causing even one fatality a year. The engineers responsible for producing telematic systems which directly intervene with the vehicle controls must, therefore, be extremely careful to ensure that their designs are fail-safe and fault-tolerant.

In this paper the potential applications of telematics to road transport are (quite arbitrarily) divided into five areas:

- Vehicle control and driver assistance (including cooperative driving)
- Route Guidance and Route Optimisation
- Emissions control
- Fleet control
- Road pricing

There is of course overlap between some of these.

4. Vehicle control and driver assistance

Traffic densities and vehicle speeds continue to rise, and there is pressure to ensure that the roads and motorways already built are used as efficiently as possible. For these reasons it is likely that electronic systems will be used to reduce the load on the driver:

- Automatic speed (cruise) control is already common.
- Inter-vehicle data transfer has been used experimentally to control vehicle separation.
- Radar systems linked to cruise control (with the aim of managing vehicle separation) have been demonstrated. Some versions just produce a warning noise, others automatically close the throttle and apply the brakes.
- Image-processing systems can be used to aid - or take over - steering.

Most research projects have used a radar sensor, but there are alternatives such as ultra-sound or optical systems. Ultrasound technology has some attractions: it is cheap, intrinsically safe, and does not raise the level of the electromagnetic background. However its performance to date has not been as good as that of the best radar sensors. Optical (video or laser) systems are potentially very high resolution, but are more expensive and suffer more than radar from dirt or adverse weather conditions.

Convoy driving using specially-designated highway lanes is currently under active consideration as a means of improving the traffic density and usage of motorways. The aim is to improve the traffic flow and density through the use of vehicle-to-vehicle communications.

Convoys of electronically linked vehicles travelling with reduced headways are expected to significantly improve road usage and reduce fuel consumption, journey time and driver stress. Convoy driving systems are the logical extension of the intelligent cruise control (ICC) systems now being deployed, although it is likely that all vehicle-based range sensing (e.g. radar) systems will have to be backed up by vehicle-vehicle data communications links to ensure that convoy driving is fail-safe.

In theory, elimination of at least part the human response from the control loop should allow vehicles to travel closer together than at present, (headway reduction), with improved safety. The introduction of telematics systems such as ICC [1] brings the advent of convoy driving much closer. However, so far insufficient research has been undertaken into the safety of convoy driving in the event of full or partial systems or mechanical failure on one of the convoy vehicles.

5. Route Guidance and Route Optimisation.
Road navigation or guidance systems are already beginning to appear and will become common in the near future. These can range from devices which warn of traffic congestion ahead, such as the "Trafficmaster" system, to devices such as SOCRATES (produced by Phillips) which guide the driver through a journey. It is only a matter of time before these different systems are combined, giving a device which provides advice on the optimal route, taking account of traffic conditions ahead.

Currently, the driver interface for navigation systems usually consists of a small screen on the dashboard. The trouble with this is that it is distracting for the driver, who has to look away from the road and focus on a screen. An alternative display has been also been tried, where the information is projected onto the windscreen. This is called a head-up display - a similar system is used by the pilots of fighter planes. Head-up displays can also be used together with a "night sight" - or thermal imaging system. However, it is unlikely that such systems will be fitted to production vehicles, although they may well be used for the emergency services such as the police, fire or ambulance.

6.Emissions
Once an infrastructure for telematics has been established, probably based on roadside radio beacons, it could be used to control vehicle emissions. There is no technical reason why the response and performance of individual vehicles should not be modified in the light of prevailing environmental considerations. For example, hybrid vehicles could be switched to zero-emission mode on entering a city centre. Alternatively, the engine management strategy of an internal-combustion engined vehicle might be adjusted to reduce emissions, at the cost of some reduction in performance, in particular weather conditions.

7. fleet management.
Effective management of a vehicle fleet can make a very large contribution to reducing its fuel usage, and hence to the costs and environmental impact of transporting goods and people.

Fleet management systems (such as the GEC-Marconi STARTRACK) consist of automatic vehicle location (AVL) devices, in which a fleet of vehicles (such as delivery vehicles) are equipped with systems which periodically report their speed and position to a base station computer.

A further application of telematics for controlling freight movement involves placing small microwave transponders (known as "tags") on vehicles or freight containers. The tags are interrogated by microwave radio as they enter the field of view of a fixed microwave beacon (tagging systems are essentially line-of-sight). Tagging systems may also be used for

- Automatic vehicle/container identification and location
- Traffic light priority control (e.g. for emergency vehicles, or public transport)
- Road pricing/tolling
- Access control for car parks
- Security (protection of unattended vehicles etc)

8. Road Pricing.

Road pricing is an effective method for managing road traffic demand, protecting the environment, and raising additional revenue for highway and infrastructure maintenance and improvement. It is firmly on the political agenda of most developed countries, including the UK. If (as seems increasingly likely) some form of road usage control is introduced, it will be essential that the technology used does not become an actual or perceived source of inconvenience. It is clear that in many cases the collection of roadside charges through 1950s-style "stop and pay" manually-operated toll plazas will be unacceptable due to the sheer size of the plazas required, as well as the disruption to traffic.

So far most automatic debiting systems (such as those in place on the Severn bridges and the Dartford crossing) have been developed for the conventional road-tolling market. They operate by passing a fixed identity code to the roadside system, which is used to identify a customer whose account (maintained at a central location) is debited for the amount of road-use recorded. This mode of operation is adequate for the purpose described, but is not capable of very much additional functionality: in general such systems cannot support

- non-stop high-speed tolling
- car park management and charging
- variable pricing (with time-of-day etc)
- international currency transactions

A number of spin-off benefits are likely to arise from the introduction of road pricing technology. By 2020, most vehicles will probably be fitted with a touch-screen, keypad or voice-activated control system. Connecting such systems to the transponder will enable the provision of, for example, route guidance and traffic information (at a price!). Once a reliable, two-way highway-vehicle data link has been established and becomes near-universal, many other services will become possible, including:

- Route guidance
- Route optimisation, based on forecasts of traffic density
- Provision of local/regional information, such as tourist attractions, accommodation etc
- Booking services for hotels, sports venues etc

- Internet access?

9. Conclusions

In conclusion therefore, vehicle telematics has the potential to radically change the way in which we use and are affected by road transport. Undoubtedly some of our current freedom to make unfettered use of the road network will be lost - but we should benefit from reduced journey times, less stressful driving conditions, and reduced environmental impacts.

The technology required to achieve many of the things I have described in this presentation already exists. The main obstacles to its widespread introduction are human, rather than engineering, and include:

- Acceptance by society.
- Evolution of a design culture and methodology aimed at producing fault-tolerance and fail-safe behaviour in telematic systems.
- Evolution of suitable man-machine interfaces to avoid overloading the driver with messages.
- Resolution of legal issues (e.g. manufacturers vs. Highway authorities liability in the event of breakdown). Can drivers still be prosecuted for driving "without due care and attention"?
- Phasing out of the bulk of the existing vehicle fleet - or retrofitting with telematic systems?

10. References.
1. *House of Commons Select Committee on Transport Report on Air Pollution,* 1993-4 Parliamentary session.

2. *Transport and the Environment.* Royal Commission on Environmental Pollution, HMSO command 2674.

3. *The Strategic Deployment of Advanced Transport Telematics.* Centre for Exploitation of Science and Technology, October 1993.

ent aeroengine control system – a stage in evolution

BORRADAILE
d of Performance and Control Systems, Rolls-Royce Commercial Aero Engines Limited, Derby, UK

Synopsis

The early jet engines had relatively primitive control systems, but as more efficient components and cycles were introduced, the complexity of the control system increased rapidly. In the last 10 years the use of Full Authority Digital Electronic Controls has become widespread, which has provided the opportunity to introduce sophisticated control algorithms and fault accommodation. The need for increased reliability and for still more functionality to support engine features such as low emissions combustors will ensure that controls technology will play a key role in the development of the next generation of aircraft powerplants.

Introduction

The jet engine has come a long way since Sir Frank Whittle's early experiments, but that progress is usually described in terms of aerodynamic refinement and materials capability. However the control system has also contributed to that development, its ever increasing functionality enabling the designer to adopt more efficient cycles and more finely tuned turbomachinery. In addition it has provided the means for a more capable and user friendly interface with both the airframe and the flight crew, and more recent systems, such as the Full Authority Digital Engine Control used on the Trent, have extensive diagnostic and fault tolerant features.

Hydromechanical Controls

Early engines had low overall pressure ratios and British designs used centrifugal compressors which typically have generous stability margins. Hence the functionality of the control system was very simple and provided little more than an altitude compensated fuel tap. . The basis of the fuel system was the variable displacement piston fuel pump, which had its origins in automotive transmissions, together with a barometric pressure control and a shutoff cock (Fig 1). At an altitude the fuel flow was proportional to throttle position so the pilot was able to impose quite fierce transients on the engine. The compression system could accommodate the working line rise during rapid acceleration, but it would have been quite easy either to over-temperature the engine on accelaeration or to cause a flame out during deceleration.

The drive for reduced frontal area, higher pressure ratio and better isentropic efficiency led to the introduction of axial flow compressors, but these had much reduced stability margins. Variable stagger stators and inter-stage bleed valves were introduced as "crutches", scheduled against corrected speed, but it was still necessary to control acceleration and deceleration rate to limit the working line and fuel/air ratio excursions. The simple pressure control of the fuel delivery was no longer adequate and flow control was added by adding a feed back loop to keep the pressure drop across the throttle valve constant. Fuel supply pressure variations were leading to cavitation erosion of the main fuel pump, so an engine driven backing pump was added to provide adequate inlet pressure. Over temperature protection was provided by a magnetic amplifier and suddenly the control system was no longer simple!

The next evolutionary step was the bypass engine, again driven by the need for reduced fuel consumption. Initially the control functionality was not greatly increased for low bypass ratio engines, but in the mid 1960s new requirements started to emerge. Much larger aircraft were being designed and while reduced fuel consumption remained as key an objective as ever, environmental issues such as noise and emissions were now becoming certification, and marketing, issues. This led to a major change to the engine configuration, with the advent of the high bypass ratio engine, with its single stage fan, small high pressure ratio gas generator and separate bypass and core exhaust systems.

In general high bypass engines achieve their highest corrected speeds at high altitude, implying that the throttle would have to be pushed forward during climb. Also flat rating of thrust with ambient temperature below typically ISA + 15°C for takeoff and ISA + 10°C for climb became standard practice at this time. Since pressure ratio is constant along the flat rating and speeds are not, pressure ratio was chosen as the primary fuel control parameter, with a schedule against inlet pressure to provide constant throttle position during climb and a consistent take off throttle position over a wide range of altitudes and day temperatures; it had the secondary advantage of maintaining thrust through temperature inversions and limiting overboost on cold days. The fuel control unit also contained the acceleration and deceleration control (based on HP compressor pressure ratio), maximum and minimum fuel stops and maximum and minimum HP compressor delivery pressure, the latter provided to maintain adequate feed pressure to the aircraft Environmental Control System during descent from high altitude. These units were very complex hydomechanical masterpieces, but they required several adjustments on installation and were susceptible to contamination and drift. Fig 2 summarises the major steps in hydromechanical control development.

When two crew cockpits were introduced in the early 1980's, flight crew workload had to be drastically reduced and the engine control system had to take over some of the functions performed by the flight engineer. This was the opportunity to launch digital electronic control systems and as at first only automatic power setting and limited crew alerting were required, the authority of the Supervisory Controller, as these units were called, could be limited to less than 50% of thrust, thereby simplifying the certification requirements. The first Rolls-Royce engines to have supervisory controls were the RB211-535C & E4 engines and once these units established a track record the benefits of digital control were quickly exploited, first on the full authority Digital Fuel Control for the RB211-524G and then on the Trent.

Trent FADEC

Although the Trent has Rolls-Royce's first civil FADEC, it is by no means the first civil full authority electronic aeroengine control system, the Britannia in the 1950s (magnetic amplifiers) and Concorde in the 1960s (analogue electronics) having set much earlier precedents (Fig 3). However digital controls offer substantially more flexibility in the choice of control system functionality and this has been widely utilised on the Trent. Fig 4 shows the major units in the Trent 700 FADEC (the Trent 800 is very similar) in diagrammatic form, and the contrast with the Derwent (Fig 1) is obvious.

The heart of the system is a dual channel Electronic Engine Control (EEC), which contains the micro-processors, pressure transducers, device drivers and I/O cards. There are separate units for power control and overspeed protection and electrical power is usually supplied by a permanent magnet alternator which also provides the HP shaft speed signal. Fuel control is provided by the Fuel Metering Unit which also contains the Shutoff Valve and the variable stator vanes are actuated by fuel driven rams. The bleed valves are pneumatically actuated via solenoid valves and the EEC also controls the starter valve, igniters, heat management system and the thrust reverser. Cockpit indication is routed through the EEC using ARINC 429 data buses, which are also used to transmit ECS and anti-ice status, and air data (total temperature, Mach number etc.). However to meet isolation requirements, each engine has its own inlet total and static pressure and total temperature sensors. Engine specific data such as rating selection and turbine temperature and pressure ratio trimmer values are stored on a data entry plug, which is individually programmed and stays with the engine if an EEC is changed.

Considerable benefit can be obtained from a digital implementation of the functions and a few examples will illustrate the point. With hydomechanical controls, the acceleration/deceleration control provided a fixed proportional over-fueling or under-fueling relative to a nominal demand. The designer was squeezed between the high fuel demand of an old engine with heavy offtake load and the low fuel demand of a new unloaded engine. A bias in one direction leads to the new engine stalling, in the other to the old engine running down. FADEC makes closed loop control of rate of change of HP shaft speed feasible, which gives a very consistent throttle response independent of engine loading or age and hence eliminates the need for maintenance adjustment. Another example is handling bleed valve scheduling. Typically the engine requires bleed over a wider range of engine speed during transients than at steady state conditions, and with FADEC it is possible to design transient detection algorithms to permit separate transient and steady state bleed valve schedules. The principal benefit is reduced fuel burn, but there can be noise advantages as well. In the sub-idle region a number of refinements to the control response have been implemented, from pre-opening the variable stator vanes during cranking to assist rotating stall drop out, through fuel dipping for stall recovery, to modifying the fuel scheduling based on engine pre-history.

However it is always a surprise how little of the EEC software is actually involved in controlling the engine. Only 20% of the code executes the control laws, the remaining 80% supports the airframe interface and the maintenance system. The latter is a very important function as it includes the fault detection, accommodation and enunciation functions. The Onboard Maintenance System (OMS) allows access to information from Built In Test Equipment (BITE) from many aircraft systems, of which FADEC is just one. The EEC manages the FADEC system BITE and has two modes of operation:

1)   Normal mode where the EEC is continuously transmitting fault reports to the aircraft;

2)   Interactive modes, where maintenance personnel can establish two way
        communication with the EEC from the flight deck and interrogate the system to
        access historical fault data or initiate functional tests on the system.

Various checks are used for fault detection (eg range, cross and model checks) and faults have been classified based on the system safety analysis (Hazard Analysis, FMECA and Markov dispatch analysis). With dual channel FADEC architecture it is possible to continue flying safely with certain faults present, using a concept known as Time Limited Dispatch. This enables the aircraft to return to its main base for maintenance action. All faults are categorised based on the safety analysis as

"NO GO" (No dispatch)
"GO IF"   (may result in operational restrictions)
"GO"   (Time Limited Dispatch, typically 10 days/150 hours)

All fault codes are stored in the EEC Non Volatile Memory and has the capability of storing up to 64 flight legs and up to 200 faults.

Future trends

The future is likely to bring further demands on control system functionality both from legislation and from market forces. Substantially improved powerplant reliability and reduced life cycle costs, emissions and noise will be demanded by customers and there will be no let up in the quest for ever decreasing fuel consumption. Translating those into control functionality is speculative, but some judgements are possible. For instance, very low emissions combustors are likely to require at least two independently controlled fuel flows, with sophisticated schedules to maintain the fuel/air ratio within tight limits. Reduced specific thrust (higher bypass ratio) tends to satisfy both noise and fuel consumption aspirations, but it is unlikely that the cost, weight and complexity of the extreme forms (which require variable geometry in either the fan or the powerplant) can be justified in the current economic climate and hence will not impact the control system in the medium term. A more likely route to fuel efficiency is increased overall pressure ratio, and better ways of controlling stability margin will be required. Power transfer between shafts is showing considerable promise as an alternative to bleed, and motor-generator and power electronics technology is steadily maturing. If motor/generators could be mounted on the engine centreline, perhaps in conjunction with electro-magnetic bearings, then the external accessory gearbox could be eliminated along with the engine oil system. Self-optimising controls which continually seek the best combination of settings to minimise fuel consumption, emissions or even noise are another possible development.

Over the years the customer requirement for reliability has become increasingly stringent, and the emerging target is zero in-flight shutdowns, aborted takeoffs and delays/cancellations. This requires a new approach to powerplant management, and the condition monitoring function in the control system will have to be expanded considerably to provide a much more comprehensive picture of the engine's health, so that preventive action can be taken in good time. This will be beyond the capability of thermocouples and pressure switches so more sophisticated "smart" sensors will be an essential element in the expanded systems of tomorrow. The fundamental reliability of the control and monitoring system will also have to be improved, and to avoid passing analogue electrical signals across wiring harnesses and connectors (a rich source of problems in current systems), distributed processing with optical signalling offers a way forward. Fault accommodation strategies will have to be developed to ensure that the increased functionality does not become itself a source of unreliability and more integration of the engine and aircraft systems will occur to avoid unnecessary duplication of functions. There are strong parallels with the automotive industry requirements and perhaps history will repeat itself, with technology being imported again from our terrestrial colleagues! However, one thing is certain; the control system will have a vital role to play in the achievement of powerplant designers' ambitions in the next century.

I would like to thank Don Yates, who provided the historical background to this paper.

Figure 1

RB 37 (DERWENT) Control System
1945

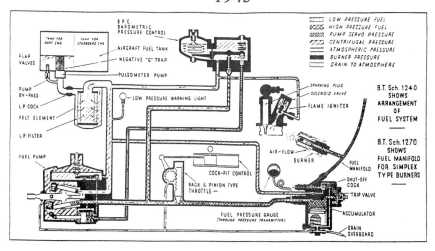

Figure 2

Evolution of Hydromechanical Controls Technology

Pressure Controllers

Fuel pumps based on car automatic transmission. Controls from barometers.	- Rods from aircraft to throttle. - Throttle barometric pressure compensated. - Variable displacement piston pump with governor.	Welland/Meteor 1944

Fuel Flow Controllers

Reduced pilot workload Increased control complexity.	- Pilot selects fuel flow. - Automatic control of accels/decels. -Altitude compensation. -Anologue electronic limiter. -Bellows and jet deflector technology.	Avon/Comet 1952 Conway/VC10 1964 Spey/BAC 111 1968

Pressure Ratio Controllers

Servo's from V1 doodlebug technology	- Pilot selects an engine pressure ratio - Compressor variable geometry (VIGV and Bleed Valves) operated by Fluidic controllers. - Analogue electronics limiter.	RB211-22/L1011 1972 RB211-524/B747 1977

Figure 3

EVOLUTION OF ELECTRONIC CONTROLS TECHNOLOGY

ANALOGUE THROTTLE CONTROL

Magnetic
amplifiers

- Single Channel control.
- DC motor pilot back-up
  control of fuel flow.

Proteus /Britannia   1956

ANALOGUE FUEL CONTROL

Transistor
analogue
technology

- Dual channel analogue, full authority.
  Thrust control.
  Primary nozzle control.
  Limiters.

Olympus/Concorde   1969

SUPERVISORY CONTROL

Microprocessors
& large scale
integrated circuits

- Hydromechanical pressure
  ratio control.
- Digital electronic trim.

RB211-535/Boeing   1980
757

DIGITAL FUEL CONTROL

⇓

- Full authority digital control of
  fuel.
  Dual channel fuel flow.
- System self monitoring for
  faults.

RB211-524G/Boeing   1988
747-400

FADEC

- Full authority digital control.
  Compressor variables.
  Bleed valves etc.
- System self monitoring for faults.

RB211-Trent/Airbus   1994
A330
Boeing 777

Figure 4

RB211-TRENT 700 CONTROL SYSTEM
1995

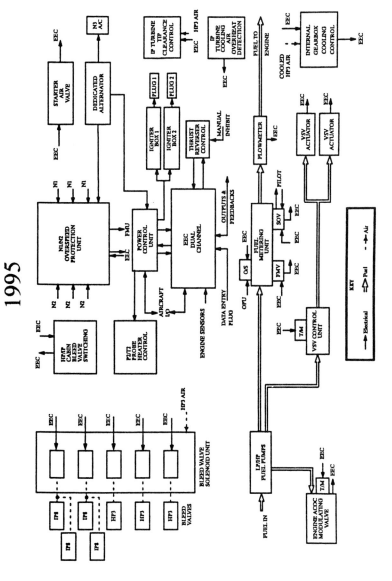

chatronics in the design of railway vehicles

essor R GOODALL MA, PhD, FIMechE, FIEE
ssor of Control Systems Engineering, Department of Electronic and Electrical Engineering, Loughborough
rsity, UK

Synopsis

For many years the design of railway vehicles has primarily been a mechanical engineering discipline. However this is set to change with the incorporation of ever-increasing amounts of electronics, the consequence of which will be a fundamental re-think of the way in which the vehicles are designed. The paper gives some examples of the impact this "Mechatronic" approach is already having upon vehicle design, identifies current and future research and development requirements, and presents some thoughts regarding the long term trends for "Mechatronic Trains".

1. INTRODUCTION

Travel is an important part of modern life, and the amount of travel increases both with the level of economic activity and with the standard of living. However, throughout the developed world the environmental unacceptability of continuing the kind of expansion which has been seen in the road network over the last two or three decades in order to cater for the ever-increasing demands for travel is well established. The automotive industry is of course looking towards methods by which higher utilisation of the existing infrastructure can be achieved (1), but it is certain that more use of public transport such a railways would make a great contribution to providing sustainable methods of transport for the future. Most people readily accept this role which railways could play in contributing to the quality of life in society, but in examining the role of technology it is important to consider the issues which affect the use of railways in more detail.

Firstly, there is cost, in particular the relative cost between rail and road. In fact the cost of rail travel, typically 25p per mile for a normal standard class journey, is currently cheaper than the true cost of driving a car, usually calculated to be around 30-35p per mile for a family saloon. However, even with government subsidies, a railway system is at a serious economic disadvantage compared with privately-owned vehicles in which many of the costs (e.g. for purchase, insurance, road tax, maintenance, etc.) are paid in advance, because the choice is often dictated by the fact that the marginal or apparent cost of travelling by road, primarily the fuel cost, is very much lower. Any cost reductions through technological advances would therefore be of great benefit to the railways.

Next, there is speed of travel. In the UK substantial decreases in journey time came with the introduction of 125mph High Speed trains during the 1970s. These have now spread over much of the intercity network, but the speed increases since that time have been rather modest (unlike other countries which now operate at 170mph and higher). Nevertheless, on journeys for which there is no direct rail connection, the sheer speed of railway offers a significant competitive advantage over road travel. The continuing worldwide quest for higher speed is primarily driven by the railways' need to compete with air travel, although there is obviously also a benefit in making less direct journeys more attractive compared to road travel.

Both cost and speed are quantifiable issues, but the third issue, that of convenience, is essentially qualitative and hence rather more difficult to pin down. The door-to-door capability of the private car is such a valuable feature that, even with the high speed of modern railways, a train journey has to be relatively straightforward (i.e. very few changes) to result in reduced journey time. In many senses this is achieving the same thing as increased operating speed, but this is only part of the story. The inconvenience of having to change trains, the need to plan out the journey, is a most important consideration for a significant proportion of the travelling public, and may well be one of the most profound barriers to really widespread use of railways.

There are of course other issues such as reliability, provision of accurate up-to-date information, cleanliness and comfort of the trains, but none of these relies so strongly upon advances in vehicle technology as the other three issues which have been identified - speed, cost and convenience. Although railways provide a rich source of challenging problems for today's engineers, technology for its own sake is no use for the competitive environment in which today's railways are operating: it must address these key issues.

2. THE PROGRESS OF RAILWAY VEHICLE DESIGN

It is very interesting to consider the evolution of railway vehicle design from a historical point of view, in particular the overall mechanical design of the vehicles and their suspensions.

The first century and a quarter of railway development was characterised by a largely empirical approach to suspension design. This "Empirical Design Period" extended throughout the first half of the 20th Century, resulting in a number of very successful vehicle designs which have been used until fairly recently. This is not suggesting there were no calculations carried out in the design, rather that the parameters which determined the suspension's fundamental performance (e.g. the speed at which the vehicle could run, how well it went round curves and how comfortable it was to ride in) were derived from experience and experiment.

In the 1960s the science of vehicle dynamics came properly of age, and what can be thought of as the "Analytical Design Period" began. Although some of the analytical ingredients had been known since the early 1900s, it wasn't until the 1960s that all the bits and pieces were put together and effective analysis became possible (2). At the same time computers became available to analyse and predict the performance of complete railway vehicles, which are still one of the most complex dynamic systems in engineering and without computer analysis would remain intractable to analyse in detail. There had of course been a number of fascinating, ingenious and highly intuitive designs prior to this time, but it's notable that the new understanding has enabled much higher operating speeds than had previously been thought possible, and has facilitated substantial innovation in the mechanical design of vehicles (3).

During the last ten years there has been a rapid increase in the amount of electronic control within rail vehicles. What is particularly significant is that this is now starting to find important roles within the suspension systems, giving a completely new perspective to what has been principally a mechanical engineering discipline for more than a century and a half, and heralding a new "Mechatronic Design Period" which will fully exploit both mechanical and electronic engineering skills in unison.

It is important to be very clear what this word "Mechatronics" means (4). Electronic systems are everywhere around us in today's world, but most of the devices with which we are familiar - mobile phones, computers, televisions, videos, etc - are primarily electronic. However in the last 5-10 years there has been a rapid increase in the amount of electronics within products which have previously been primarily mechanical. The engine of a modern automobile is a particularly good example, because the all-pervasive nature of electronic control within the engine is evident to anybody who ten years ago was accustomed to being able to service their car with a couple of spanners, a screwdriver and a set of feeler gauges. While the basic concept of the reciprocating internal combustion engine is more or less as it has always been, the improvements in terms of increased power to weight ratio, reduced fuel consumption, lower noxious emissions, etc. are primarily a consequence of the synergy between the basic mechanical system and the electronics embedded within the engine (5).

This then, is Mechatronics: the process of taking advantage of the capabilities of modern electronic control to realise the full potential of the mechanical system (6).

3. CURRENT USE OF ELECTRONIC CONTROLS IN RAILWAY VEHICLE SUSPENSION SYSTEMS

Broadly speaking the suspensions of a modern railway vehicle can be subdivided into the secondary suspension between the bogies and the body (mainly concerned with providing a good ride quality), and the primary suspension between the wheelset and the bogies (mainly concerned with giving stable running and good curving performance).

The main area in which electronic control is currently being applied is for tilting trains, which are increasingly becoming accepted as standard for achieving higher speeds without needing major changes to the track. Tilting essentially applies electronic control to the secondary roll suspension of the vehicle. The idea is to lean the vehicles inwards on curves, thereby reducing the curving acceleration felt by the passengers and enabling higher speeds to be achieved. Strictly this can be accomplished purely mechanically (i.e. without electronic control). However the performance of such schemes is limited, and most tilting trains now use some sort of tilting mechanism fitted to the bogie combined with hydraulic actuators to give rotation of the vehicle body about the desired tilt centre. Early intuitively-derived control laws, based primarily upon acceleration measurements, proved not to give sufficient performance in practice, and modern tilt controllers take full advantage of digital processing and utilise additional devices such as gyroscopes, displacement sensors, etc.

A number of European countries have developed tilting trains: Italy with the Pendolino concept, Sweden with its X2000 trains, Spain with the Talgo, and more recently Germany with the 611/2 multiple units and Switzerland with a concept developed by SIG (7). Further afield Bombardier make a tilting train in N America and there are a number of Japanese systems either in operation or at an experimental stage (8). Although the natural choice of actuation for mechanical engineers is hydraulic, it is now being recognised that electro-mechanical actuators which take advantage of cheap, efficient and compact electronic power control devices have a number of benefits to offer. A number of European manufacturers are adopting this approach, which is significant from the mechatronic point of view because it is an indication of engineers starting to undertake design in an inter-disciplinary manner.

The commercial rationale for tilt is that gives shorter journey times, and therefore brings in more revenue by attracting more passengers. This is offset by the extra cost of the tilt system - often estimated to be around 10% of the vehicle cost, although in the longer term many observers feel that the additional cost of tilt will be marginal. The significance however is that tilt has now established a precedent for the use of electronic control within the basic mechanical structure of railway vehicles operating in service, paving the way for a more general use of Mechatronics within vehicle suspensions.

Apart from tilt, the only active secondary suspension known to be running regularly in service is a relatively simple active lateral suspension using pneumatic actuators which is associated with the tilt system on the Fiat Pendolino trains, but there are also a number of successful examples of full scale experimental implementations. More detail of the systems which are mentioned in the following paragraphs are included in reference (9). One of the earliest major experimental study was undertaken by British Rail in the early 1980s. This covered both lateral and vertical secondary suspensions using a variety of actuator technologies, and was the first practical demonstration that significant improvements in ride quality could be achieved, with improvements of around 35%-50% during mainline testing.

Since then Adtranz, one of Europe's major railway manufacturers, has tested at least two active suspension devices: in the UK, what they call their "active hold-off device", which is designed to improve ride quality simply by holding the suspension away from the limits of suspension travel on curves, and therefore improve the ride quality; in Sweden they have studied an electronically-controlled lateral damper for use on their X2000 tilting train. Other people have studied the use of semi-active devices, but they have usually modified a normal hydraulic damper whereas the ABB approach is rather novel involving a miniature electronically-controlled brake, another example of an innovative inter-disciplinary approach.

A very significant development is being conducted by Siemens SGP in Austria on an experimental vehicle (10). They have semi-active secondary vertical dampers, full-active pneumatic actuators to keep the lateral suspension centred on curves, and electro-mechanical tilt actuators (mentioned before in connection with trends in tilting trains). What is particularly interesting about this development is the way the control systems have been integrated, and it represents the most advanced Mechatronic implementation which is currently available.

In Japan there has been a lot of active suspension work, much of which appears to be closely related. Some of this has been directed towards achieving 350 km/hr operation for the Shinkansen trains, but also they have a three car "Try-z" test train for evaluating three different active suspensions concepts for conventional trains (8).

The current use of electronics within primary suspensions is much more limited, even at an experimental level, although there is an example of a Rotterdam tram in which the axles are steered under electronic control enabling extremely tight curves to be negotiated.

The commercial rationale for these more general applications of electronic control is less well developed than for tilt: for the secondary suspensions it is related to ensuring a sufficient ride quality at higher speeds, for the primary suspensions to reducing wheel and rail wear (a significant factor in the overall cost of maintaining a railway).

4. FUTURE POSSIBILITIES

To an extent the applications just described are not mechatronic in the purest sense of the word, because the integration of the electronic and mechanical components is not highly developed, and so it is useful to consider what might emerge if full mechatronic integration can be achieved.

Firstly, much lighter weight constructions are possible. Although materials and manufacturing techniques are available, their use is limited by the fact that problems emerge with the flexibility of the vehicle structures; also, achieving a good ride quality is generally more difficult with lighter vehicles, but both difficulties can be overcome by incorporating active elements into the mechanical scheme and designing the system as a whole.

Secondly, new and simpler mechanical schemes are possible. The bogies of today's railway vehicles add significantly to their weight and cost. If however the conventional coned wheelsets, which essentially provide steering through mechanical means (not always very effectively), are replaced by electronically-steered wheels then the bogie becomes redundant leading to large benefits in terms of mechanical simplicity, weight, space and cost - just the kind of influence which could really assist railways in their competition with road travel.

Such benefits can only be achieved with a high degree of electronic control which offers the levels of safety and reliability to which we have been accustomed with the purely mechanical system, a not inconsiderable task, but similar to what has been achieved in the aerospace industry with the introduction of fly-by-wire controls. The key to achieving these technological advances is the ability to bring together the mechanical and electronic control engineering techniques with systems engineering issues such as safety, reliability, maintainability and tolerance to failures of individual components within the system.

It is interesting to note that a number of functions are provided through the contact between the wheels and the rails - traction and braking as well as suspension and guidance - and in the long term it is possible to conceive of "intelligent wheels" which use a full implementation of mechatronic principles in the design of the running gear, using independent wheels in which electronics provides an integrated system to control traction, braking, guidance and suspension forces in an optimal manner.

So far the paper has concentrated upon approaches by which Mechatronics can help to tackle two of the issues identified in the Introduction: speed and cost. The third issue which was mentioned was convenience, so how can Mechatronics help this issue?

To answer this, consider first the ideas for active guidance described above: what is interesting is that in principle this offers the possibility to switch routes from on-board the vehicle, with continuous track in both directions through the switch or points and using the vehicle guidance sensors to follow whichever direction is selected. Combine this with trains in which the individual vehicles are coupled through electronic control rather than using a mechanical linkage, and suddenly we have provided the ability for individual vehicles to come together from different origins to form a train on the main intercity routes, diverging at the end to go off to different destinations. Clearly there would be major changes to the train control and signalling system, but at least the vehicle technology is possible. Interestingly, such an approach would have railways operating in a similar manner to that envisaged for Intelligent Highway Systems of the future: individual vehicles coming together to form automatically-controlled platoons or convoys on the major highways, spreading out to individual destinations at the end.

5. CONCLUSION

The paper has exposed a number of opportunities which might come about as a consequence of extensive use of electronic controls within "Mechatronic trains". Some of these such as tilting trains are already with us; other ideas such as active guidance rely upon considerable research and development before they could be applied in practice; and the ideas portrayed at the end of Section 4 are highly speculative but could give a fundamental change in the way railways operate.

What follows is a prediction for "generations" of railway vehicles which will emerge during the Mechatronic Design Period referred to in Section 2:

A first generation of Mechatronic Trains with just one suspension function under control (e.g. tilt only), having limited performance and a relatively low level of integration with other functions.

A second generation with a number of functions being controlled (e.g. lateral and tilt), having a higher degree of integration and systems designed for high reliability and tolerance to failures.

A third generation with a further increase in functionality, highly integrated with other vehicle systems such as traction and braking and linked in with facilities such as track profile databases, satellite positioning systems, etc.

Time will tell how accurate these predictions are. However one thing seems certain: electronic controls within railway vehicles are here to stay, and experience with other products and systems tells us that their use is likely to increase. Railways will continue to have an important role in society, and Mechatronics is one aspect of advancing technology which will have a significant role to part in their future development.

REFERENCES

1.   S Shladover, "Review of the state of development of Advanced Vehicle Control Systems" Vehicle System Dynamics, Vol 24, pp 551-595, 1995.

2.   A H Wickens, "Steering and dynamic stability of railway vehicles" Vehicle System Dynamics, Vol 5, pp 15-46, 1975.

3.   R Illingworth and M G Pollard, "The use of steering axle suspensions to reduce wheel and rail wear in curves", Proc IMechE, Vol 196, pp 379-385, 1982.

4.   R Comerford, "Mecha....what?, IEEE Spectrum, pp 49, Aug. 1994.

5.   J De Cirro and M Ross, "Improving automotive efficiency", Scientific American, Vol 221, No 6, pp 30-35, Dec 1994.

6.   J A Rietdijk, "Ten propositions on Mechatronics", Mechatronic Systems Engineering 1, pp 9-10, 1990, Kluwer Acad Publishing.

7.   Proc IMechE Seminar S479 "Tilting Trains for the UK?", London, Jan 1997.

8.   K Sasaki et al, "Active tilting control of series E991 e.m.u. experimental trains", Proc STech96, Paper C514/055/96, IMechE, 1996.

9.   R M Goodall, "Active Railway Suspensions: Implementation status and Technological Trends", Proc 15th IAVSD Symp, Budapest, Aug 1997.

10.   A Stribersky, "On dynamic analysis of Rail Vehicles with electronically-controlled suspensions", Proc 14th IAVSD Symp, pp 614-628, Ann Arbor, USA, 1995.

Robotics: where we have been, where we are now, and where we may be going

Professor J M BRADY BSc, MSc, MA, PhD, MIEEE, FEng, FIEE
Professor of Engineering Science, Department of Engineering Science, University of Oxford, UK

1   Synopsis

The use of robots has already had a profound effect on manufacturing, not least in contributing to the manufacture of cars that have a warranty of years and tens of thousands of miles, and in the manufacture of computer chips, an area in which humans are environmental disasters! Robots bring together sensors, actuators, control computers, and smart software that can plan assemblies and collision-free paths. We offer a non-science fiction view of the future real applications of robots in a wide variety of domains, from surgery to agriculture to the uses of robots in the home.

2   A selective bit of history

The word "robot" derives from the Czech word for "worker", and was first used in Karel Capek's futuristic play about automation *Rossum's Universal Robots*. Of course, the idea of a robot is ancient, indeed one can trace it back to the earliest times in almost all literatures. Over the past one hundred years, robots have figured large in the emerging science fiction literature. Several, including Isaac Asimov's *I Robot*, Arthur C. Clarke's *Hal* in *2001*, and the irrepressible *R2D2, C3PO* in George Lucas' *Star Wars* films, have caught the popular imagination. In science fiction, robots can locomote (legged for *C3PO*, wheels for *R2D2*), carry out complex manipulations, sense their environment (often using sensory modalities not available to humans), and, most crucially, are possessed of super-human powers of memory and intellect. The real world of robots is entirely different.

The earliest applications of "robots" in industry were pick-and-place tasks in the 1960s, often in environments that are hostile to humans, typical of forging and casting. A robot siezes a billet from a set position, thrusts it into a furnace for a preset period, then transfers it to a forge where it can be hammered into the desired shape. The requirements of such an application dictate the structure of the "robot": a simple end-effector can grasp cylindrical billets; and two degrees-of-freedom (dof) suffice for moving the billet to the furnace and then to the forge. Typically, these consist of a rotary azimuthal joint and a prismatic joint to move the billet in and out of the furnace. The mechanical design aims at end-point accuracy and repeatability, load-carrying capacity, and stiffness. The robot has no sensing, other than internal joint position sensors, so if a billet isn't presented in the right place, at the right time, the robot will still go through its preset motion, but to little effect. "Programming" can be carried out with a standard 1960s control technique such as ladder diagrams. This is a far cry from *R2D2*!

The "great leap forwards" in the development of industrial robotics stemmed from their widespread utilisation in the manufacture of cars, beginning in the 1970s. Some readers may recall the generally awful quality of car manufacture in the early 1970s: one hoped that one did not buy a "friday afternoon" car, for it would surely break down after a few days. A really well-made car might go as long as six months without having problems! Nowadays, all new mid-range family cars come with a six year/sixty thousand mile warranty! This quite extraordinary transformation has been brought about by manufacturing processes that guarantee uniformly high quality of product, and one important component of such processes is the use of robots. More importantly, cars have been *redesigned* to enable them to be constructed using robots. For example, Ford designed the *Sierra* so that 88% of all spot welds could be carried out by robots or by multi-welders, leaving only 503 to be applied manually, compared with 3508 on the *Cortina*. The two applications of robots that are dominant today are spot welding and spray painting, together accounting for over 40% of robot sales. Robot arms that are specially designed for these two tasks have been developed, and though spray painting robot arms move more smoothly than spot welding robots, and are rate-rather than position-controlled, they have very a similar kinematic structure: an open-link kinematic chain that crudely resembles the human arm. Sometimes the elbow joint is enclosed in a four-bar linkage for extra stiffness, but this makes very little difference to the overall structure. "Programming" a spray painting or spot welding robot is almost always by record-and-play-back: a human spray painter holds a paint gun that is mounted on a spray-painting robot, driven passively, and executes a series of painting runs. The joint coordinates are sampled regularly (say at 30 Hz), and recorded on a cassette tape. The single or set of runs is generalised into a trajectory: a set of joint coordinate demand points that are passed sequentially to a rate/position controller. Changing to a new model requires re-programming of the trajectory. Just like the forging robot, spray painting and spot welding robots tend to have only internal joint position sensors: if a car to be welded is presented in the wrong place, or if the clearance between the metal sheets to be joined is not within the limits required by the spot welding torch, the parts may go unwelded, though the robot will have no knowledge of that fact!

We noted above that until the mid-1980s most robot arms crudely resembled human arms, in that they were predominantly open kinematic chains, with a series of single dof, rotary joints. The robot arm typically occupies a plane that can be oriented by an azimuthal joint. Within the plane of the robot arm, there is usually a shoulder, elbow, and wrist joint, terminating in a one dof end-effector whose form is application dependent. Ignoring, for the moment, the end effector, there are 3 dof plus the dof of the wrist. In general, six dof are needed to specify a position and orientation in space (say of the end-effector). For this reason, a number of 3 dof (3 roll) wrists were developed for robot arms. It was established during the early 1970s that the kinematic equations of a robot wrist could be inverted if the three rotary axes intersected in a single point. Unfortunately, the majority of 3 roll wrist designs were mechanically quite weak, leading to otherwise powerful robot arms that were wimps because of weak wrists. However, it was also realised that many industrial applications do not require the full six dof: since many involve picking objects off a planar surface, 2 or 3 translations plus one rotation often suffices. Since additional joints tend to reduce end-point stiffness and increase cost, manufacturers usually offer a 4 and a 5 dof version of an arm. It is important to realise that the resemblance of such a robot arm to the human arm is crude in the extreme! The human arm is powered by coactuating tendons that outperform any extant motor, except in power. An individual human joint (eg the shoulder) may be powered by as many as 18 muscles, each of which is an exquisitely programmable engine! This is a point to which we will return later.

The appearance of micro-, and more usually mini-processors, during the 1970s meant that robot arms became, for the first time, genuinely programmable, in the sense that manufacturers provided robot programming languages, originally developed in university laboratories. Typical of the genre was the VAL programming language, which resembled BASIC to which had been added homogeneous coordinates to define static or dynamic coordinate frames affixed to points in the work space or on the robot. Perhaps the most important command in VAL was *move* X_g, which moved the robot arm along an automatically-computed trajectory to a specified end-point X_g, and *moves* X_g, which did the same along a straight line in world space (ie not in joint space). VAL was a major step forward in programmability, though the fact that it was hard, if not impossible, to interface sensors to the arm, and to close the loop around sensory information meant that it was also extraordinarily frustrating to use, at least for more advanced applications.

Early in the 1980s it was realised that so far as the emerging application of robots to semiconductor (board level) assembly was concerned, the human arm-like kinematic structure could be bettered. This led to the SCARA structure, nowadays used almost exclusively for board assembly. The SCARA consists of two rotary joints that move their end point in a horizontal plane, the end point consisting of a prismatic joint that moves vertically up and down, so that DIPs can be inserted rapidly in their slots on printed-circuit boards. Semiconductor manufacturer tends to be based around manufacturing cells that are populated with programmable devices such as x-y tables that can translate a board at high speed. One can go much further with semiconductor manufacture than board stuffing. Indeed, since humans are a major environmental disaster for semiconductor chip making, there is every reason to do so. This led Victor Scheinman, original designer of the Unimation *PUMA* arm, to propose in 1987 the concept of *robotworld*[1], in which the design of a system, in particular the kinds of sensors to be used, and the scale at which operations are carried out, is developed, in a generic way, from the application and not from preconceptions that lead to making robot arms in an approximate image of human sensors and arms. To date, this idea has essentially remained a laboratory concept; though there has been some quite encouraging progress recently (see Ish-Shalom's article in [2]). It seems certain that this idea will be developed to become a commercial proposition in the near future.

Though most industrial effort has been expended on robot arms, there has also been a great deal of progress in making robot vehicles. These were developed originally for parts transfer in factories, in cases that are difficult or costly for conveyors, for example of relatively heavy parts, or in cases where a part could follow a number of different paths through a manufacturing plant. This latter is typical of job-shops, as modeled in discrete-event systems. Up to the present, the majority of vehicles have been wire-guided and can follow only a preset path, most often with no branches. This lack of flexibility pays for the increase in end-point accuracy relative to so called "free-ranging" vehicles, to which we return below. Free-ranging requires two distinct abilities: (i) the robot should know accurately its position at a set of positions that is far larger than can be reached by wire guidance; and (ii) the robot vehicle should be able to detect the presence of obstacles and plan collision-avoiding routes around them to get to the original goal. As regards the first of these, in the mid-1980s, GEC Electrical Products developed a position sensor based on a retro-reflecting infrared laser that estimated the bearings of a number of bar-coded targets placed in a workplace, from which, after the positions of the targets had been surveyed, the robot could position itself to an accuracy better than a half a centimetre over a range of 15 metres. Substantial developments of the idea by the start-up company Guidance and Control Systems (GCS) have extended this to better accuracy over 150 metres! Significantly, this has involved two separate developments. First, a complete mechanical and optical redesign of the rotating laser scanner, and second, smarter processing built into the scanner. The targets are no longer bar-coded, because the thin stripes of the bar code, compounded with the point-spread function of the optics, limits the effective range. GCS use a single wide stripe and kalman filtering to determine the robot's position. The signal processing and filtering is carried out inside the scanner, using a LICA (Local Intelligent Control Agent) processor developed by the author and Dr. Huosheng Hu at Oxford University on a project supported by the EPSRC. This is a typical example of mechatronics pervading robotics: the replacement of a simple opto-mechanical component by an integrated system that comprises optics, mechanics, signal processing, and computation. The LICA technology is now available from GCS in a number of products, and for a number of applications, under license from Oxford University. Together with a number of other positioning sensors such as radar, as well as GPS (or differential GPS), for many applications ego-positioning can be considered a solved problem. The second problem motivated our research at Oxford and will be discussed briefly in the next section. There are now hundreds of robot vehicles fielded in industry, for parts transfer in manufacturing, but also in applications such as moving containers around busy ports and controlling the motions of mining machinery.

A final application area of robotics in manufacturing concerns parts acquisition and/or inspection. Of course, vibratory bowl feeders have enjoyed considerable success, though there is a quite strict limit to the size and composition of objects that can be fed in a known position and orientation from a bowl. The promising development of sensor-less parts feeding and manipulation is described in the next section. Inspection has typically used binary vision, a technology that has become more attractive as (i) memory and processing prices tumbled, and, more importantly (ii) algorithms were developed that actually work in practice! Key to (ii) is often careful control of the arrangement of the lighting. A good deal of *ad hoc* knowledge is now available about how to achieve this.

The emphasis above has been on manufacturing, since that is what has driven the development of practical robotics. However, we note briefly a number of other specialised sectors in which the design and use of robots has become highly developed. First, nuclear processing offers a range of challenges to developing robots that have to work in a highly hostile environment. BNFL and AEA Technology have become world leaders at developing nuclear-hardened robot arms and, equally importantly, advanced teleoperation schemes that enable a operator to control the arm remotely. For example, schemes for force feedback and position control have been developed [3]. Related tele-operated devices have been developed for subsea exploration and rescue. The locating of the *Titanic* was a spectacularly successful case of this . At the end of 1993, for the first time in the relatively short history of space travel, a small, multisensory robot performed a number of tasks on board the shuttle *Columbia*. Gerd Hirzinger and his colleagues at the DLR near Munich developed the *ROTEX* system successfully opened and closed connector plugs (bayonet fitted), assembled structures from single parts, and caught a free-floating object. To do this, the robot was pre-programmed (and re--programmed from the ground), and teleoperated by the astronauts using a control ball and a stereo-TV monitor.

We close this section with a comment about "definitions". The Robot Institute of America famously, if long windedly, defined "a robot is a reprogrammable, multi-functional manipulator designed to move material, parts, tools, or specialised devices, through variable programmed motions for the performance of a variety of tasks". Re-markably, the definition makes no mention of sensing, and is generally a reflection on the applications that had been fielded by the mid 1980s. Fifteen years ago, the author offered the slogan that "robotics is the intelligent connection of perception to action", while David Grossman, then head of IBM's Yorktown Heights robotics lab-oratory quipped that "a robot is a surprisingly animate machine"! Joking aside, an important aspect of the two latter definitions is that they evolve as understanding of robots evolves: today's intelligent robot is tomorrow's unremarkable machine. Consider as an example the automatic disc changer in a modern CD player: it has a 2 dof manip-ulator arm, simple sensing, and follows a programmed trajectory. Not very remarkable, but considerably smarter than almost all robots were twenty years ago!

3 Where robotics research is now

Whereas the previous and following sections emphasise applications and markets, this section is largely concerned with technology. We briefly sketch a few recent develop-ments in the emerging science and technology of robotics. The reader wanting more detail might start by perusing the articles in [4, 2, 5], or by emailing specific questions to the author.

Development of novel kinematic structures

In the previous section we noted that wrist design is often the weakest feature of a robot arm, observed that end-effectors are often simple one dof devices, and observed that arms have largely been designed to transport the end-effector to the point of action. These limitations, among others, have been addressed in recent research. First, it is occasionally necessary that a wrist have the full three dof necessary to control the orientation of the end point, and, worse, that it be necessary that the wrist react extremely rapidly to sensed positions and forces. This has led to a number of novel wrist designs, the ET wrist developed at the University of Western Australia being one of the most advanced. The goal of the project is to shear sheep as rapidly as a human shearer yet without cutting the skin of the sheep. The problem is to "fly" a very sharp cutter a millimetre or so above the skin surface, measuring and reacting to the position of the skin surface as the sheep breathes. It is vital to maintain the axis of the follower actuator as close as possible to the normal to the computed surface of the sheep. The ET design incorporates capacitance proximity sensors and force sensors, has well-conditioned kinematic and dynamic performance throughout a large workspace without enclosing singularity cones. The wrist is light and relatively strong.

Second, a number of dextrous hands have been developed (see [4] for a list of further references). The design of such hands is a tough problem for mechatronics. Three fingers suffices to grasp an object with friction, a fourth enables regrasping without putting it down. Most dextrous hands have been tendon actuated, to enable the motors to be remote from the ends of the fingers in order to increase the range of objects that can be manipulated. Tendon material is a critical parameter of the design, if it is not to degrade or crack after a few thousand cycles. Though dextrous hand mechanisms are quite complex, several have been developed, leading in turn to serious study of grasp planning and control. One approach has been to develop hand pre-shaping based on a crude analysis of the shape of the object to be grasped. Early work on human grasping suggested that six categories of hand shape sufficed to account for the majority of objects in practice. A different approach has been to compute a representation of the shape of an object, useful for determining a grasp. For example, for a parallel pair of fingers, a symmetry set representation enables grasp configurations to be determined automatically [6]. This latter approach has the virtue of being applicable to a wide range of object shapes, though much more research is necessary.

Third, the current concentration on end-effector manipulation is in contrast to our human experience in which the whole of our arm is used, if necessary, to carry or manipulate an object. To this end, Salisbury [2] has developed the concept of whole-arm manipulation, and has developed a lightweight arm called *WHAM*, that enables non-traditional robotic gestures such as pushing, shoving, striking, cradling, cushioning, and interlink grasping. The links of *WHAM* are made of lightweight aluminium covered in dense foam to provide reasonable friction and impact characteristics. The arm uses stiff cable transmissions and brushless motors to maximise the useful workspace and to facilitate force control. A second, three dof version has been developed at Woods Hole Laboratory and operated at full ocean depths as part of the *JASON/ARGO* system.

Finally, it has been realised that reliable real-time vision systems can be constructed, particularly if the steering control of a stereo (two-eyed) camera platform is integrated with image processing. That is, not only does information gleaned from an image determine the next motion of the platform; but "deliberate" movements of the platform can be made to elicit required information. A number of high performance "intelligent" stereo platforms have been built, not least at Oxford by Dr. David Murray and his colleagues, and by the author. This has contributed to the development of a new approach known as "active vision", to which the book edited by my colleague Professor Andrew Blake [7] is a good introduction.

Sensing

Vision aside, many robot arms and vehicles use ultrasound range sensors, not least because they are cheap and give relatively rapid and accurate range readings, at least when used as proximity sensors (eg in a "safety curtain" application). Over ranges greater than about 2m, ultrasound is prone to specular reflections leading to false range readings. Leonard and Durrant-Whyte showed that environmental features, such as corners and walls, correspond to regions of constant depth (RCD), portions of the signal that in polar coordinates have constant range. By tracking RCDs over known robot motions, one can determine the type of the feature (concave corner, etc) and so build up a map for navigation. The idea has been extended to qualitative landmarks, using ideas of Artificial Intelligence. As for the GCS scanner described above, one can build a "smart" ultrasound sensor incorporating a LICA that delivers RCDs in real time! A second problem with ultrasound is that commercially available transducers tend to report the nearest echo over a broad angle, often 24 degrees. Infrared (IR) has far better angular resolution, but IR range sensors have to overcome the speed of light problem: IR travels one metre in a gigasecond, so direct measurement of time-of-flight requires very fast sampling. However, a number of range sensors have become available based on the observations (i) that range is directly proportional to phase, which can be sampled in a phase-lock loop; and (ii) that phase can be heterodyned to 68 KHz, hence computed with TTL logic. The quality of range data from such sensors is high, though so, for the moment, is the cost. It will hopefully not surprise the reader that one can build a IR-LICA that delivers not the raw IR signal but landmark features extracted from it. No single sensor delivers reliable information all of the time, so most systems comprise a distributed network of sensors whose results are fused together, most often by some variant to parametric decision analysis, though some initial forays have been made into non-parametric decision making using Dempster-Shafer theory or qualitative reasoning. We have learned how to construct networks of smart sensor "agents" that communicate with one another to determine the layout of the environment, without using a single central processor.

The previous section referred to "the promising development of sensor-less parts feeding and manipulation". Working at MIT, Matt Mason began by asking: "if a planar object, resting on a planar surface with friction, is pushed quasi-statically in a given direction, which way will the object move?". This was the basis, for a theory of planar pushing, one of whose practical implications is the use of a set of object-specific guide rails above a conveyor belt that eventually, and provably, transform an object coming along the line from its initially unknown position and orientation to a configuration that is known, from which it can be grasped by a sensor-less device. This led in turn to John Canny's idea of Reduced Intelligent-Sensor Computing (a play on acronyms of RISC) in which, starting from the goals of an application one designs simple sensors and guide rails to remove uncertainty about the configuration of an object. There are huge potential savings both in speed and cost for many applications.

Vehicles

We noted in the previous section that we have had a fruitful relationship, first with GEC Electrical Projects (now CEGELEC), subsequently with GCS. We have concentrated on developing LICA-based smart sensors for sonar, infrared, for the GCS scanner, for a lateral-effect photodiode range sensor (in collaboration with BNFL), and a stereo vision camera platform. We have also built a LICA-based control and trajectory planning system so that the second attribute mentioned in the previous section for a free-ranging vehicle, namely "the robot vehicle should be able to detect the presence of obstacles and plan collision-avoiding routes around them to get to the original goal" can be solved, at least for known workspaces. Other work on robot vehicles has included the Frait vehicle for dockside container movement, and Silsoe's work on robot selective crop spraying aimed at reducing the amount of chemical poured on to the land, work that was a collaboration with my colleague Professor Andrew Blake.

The European project PROMETHEUS brought together a number of car companies, component manufacturers, and research laboratories throughout Europe, with the aim of developing a range of products for "intelligent vehicle and highway systems" (this is the name of the subsequent US counterpart of PROMETHEUS). Though the emphasis of PROMETHEUS was on the driver-in-the-loop, it featured a number of projects that provide advanced sensory information: cruise control based on a radar, lane following, infrared image processing for night driving, and parking aids. Some of these are on course to becoming available commercially. Other related work has aimed at making driver-less cars, with "demonstrations" that often flatter to deceive. Among these are the traversal of the US by a vehicle developed at CMU, and the round-trip between Munich and Stuttgart by a vehicle developed by Professor Dickmanns of the Technical University of Munich.

Legged locomotion

There has been a considerable interest in developing legged robots. In Japan there has been no attempt to rationalise such work in terms of industrial applications, rather it has been a question of whether such systems could be built at all and if so what they might be capable of. Mostly, work has concentrated on systems that are statically stable, with a moving triangle of support and duty cycles of the legs. Six legged systems and four legged systems have been developed that can climb over obstacles. A huge six legged system was developed in the USA for the US Army, the rationale being that battles are rarely fought on highways (though in Los Angeles this is not clear). Only rarely has the problem of foot placement, based on an estimate of the perceived load-bearing quality of the terrain, been addressed, and this remains essentially an open question.

Other work in Japan, most notably by Professor Miura at Tokyo University, and in the USA by Professor Raibert at MIT, has studied biped locomotion, hence dynamic stability. Raibert and his colleagues have developed biped systems that can perform gymnastic stunts like somersaults and can leap through a hoop! More recently, Professor Inaba, working in Miura's laboratory at Tokyo University, has developed an impressive system that could, I believe, have enormous market potential. He has developed a two-legged, two-armed robot system that looks like a metre-high doll. The body is covered in touch sensors, and the "head" has a range of sensing systems, including a simple vision system. From an initial position lying on its back, Inaba's robot can sit up, then stand up, and can operate a swing!

Path planning

For all but the simplest workspaces, planning collision-avoiding paths for a moderately complex robot is a very difficult problem. One approach, that often works well in practice, though cannot guarantee that the robot will not get stuck part-way, is to regard the robot as carrying a negative charge, the goal position to be a positive charge, and obstacles negative charges, then solve the corresponding Laplace equation. There have been numerous improvements to this basic scheme, including using Maxwell's current laws, and changing the distance metric for the Laplace equation. Better, Barraquand and Latombe, working at Stanford University, have developed a probabilistic path planner that gives excellent results in reasonable time in many practical cases. Related work has been carried out for collision-avoiding path planning for car-like vehicles. The non-holonomic kinematics of cars complicates the problem considerably, and, reassuringly, one can explain mathematically why it is so hard to park a car between two others at the side of the road! It has been shown by Reeds and Shepp that the optimal path from a start configuration to the goal configuration is a spline consisting of circular arcs; unfortunately, this means that path curvature has to change discontinuously at the knot points. Fortunately, clothoid curves are a good approximation to Reeds-Shepp curves, have continuously varying curvature, and are fast to compute.

More recently [5, 8], there has been rapidly growing interest in image-guided surgery.

Particularly strking advances have been made in spinal cord surgery, and in brain surgery. In Taylor's work at IBM Yorktown Heights, a robot arm positions a NC drill near an exposed hip in order to make a precise hole for a hip implant. Path planning is performed off-line on the basis of x-ray images. The surgeon is always in the loop, controlling the computer-guided tools much like he/she would any other tool. A key problem is registering data taken before an operation (eg a brain MRI in the case of a brain tumour) with information available during the subsequent operation, so that tools can be guided precisely toward the target along a path planned in association with a surgeon. This is an active area of research, aimed at avoiding the use of stereotactic frames.

4 Where robotics may be going

It is important at the outset of this section to remember two things. First, many forecasts have been made over the past thirty years concerning future applications of robots, and they have often been either hopelessly optimistic, particularly about timetable, or just plain wrong. I am confident that all of the developments alluded to below will happen, but I am less confident about when. Largely that is because the timetable will be driven largely by market forces, and that is hard to gauge, particularly in areas where currently no market exists! Second, today's "smart" robot is tomorrow's unremarkable machine. Remember the automatic CD changer, and people movers such as the Orlyval service in Paris. It is equally important to recognise the context in which we ask where robotics may be going: the remarkable pace of developments in IT, and the equally remarkable convergence of IT and Communications. This has been well summarised in the *IT demographics* section of the UK National Foresight report on IT (available from the Stationery Office), which sets out what *we can already be sure will happen* over the next ten years. This impacts on the future of robotics since, as has been made clear in earlier sections, increasingly it is software and mechatronics, rather than mechanical design per se, that is at the heart of developments in robotics. Indeed, it is likely that one of the most fertile areas for developing the subject will be in an extension of Scheinman's *robotworld* to the level of micromachines.

However, it is unlikely that the traditional domains of robotics are going to reduce; rather one sees robotics being applied even more comprehensively in, for example, the construction and operation of cars. In terms of construction, there is still a great deal that can be achieved in automating engine construction and in inserting components in instrument panels. Perhaps this will necessitate yet another iteration of car design. In terms of car operation, as both the PROMETHEUS project and the UK Foresight project report on Transportation have shown, advanced transportation systems face a number of considerable challenges: the need for ever improved reliability, increased safety of vehicle drivers and passengers, the need to manage effectively the rapidly increasing density of all forms of transport, and the need to minimise the environmental impact. And all this at reduced cost as the cost/performance trade-off becomes ever keener in the market place. Mechatronics enables the development of "smart" sensors and actuators, while dedicated processors such as DSP chips make feasible the real-time deployment of advanced signal processing algorithms that were infeasibly slow only a couple of years ago. Novel hardware technologies such as field-programmable gate arrays (FPGA) blur the distinction between hardware and software enabling a common design approach to a whole system. Open standards, originally developed for local area networks and for workstation design, are increasingly being designed and adopted throughout manufacturing, in particular for the manufacture, but also the control of advanced vehicles. Drive and fly by wire are becoming increasingly common.

Similarly, robotics will continue to develop in the traditional areas of application that are hazardous *to* humans, such as nuclear assembly (and, more importantly in the UK, decommissioning); maintenance of offshore platforms and subsea exploration and recovery; and military applications. The latter will continue to have many dimensions, from smart weapons, through munitions loading machines in confined spaces typified by tanks, smarter bomb disposal systems, autonomous drone aircraft, sentry robots, and perhaps vehicles for triage. This timetable in this area is particularly difficult to foresee, since it clearly dependent on military budgets, which seem likely to continue to be under close scrutiny, and because ideas about likely future theatres of conflict are also under review. Some of these latter applications have civil counterparts too, for example, there may well develop a market for helicopter drones carrying sensor suites that enable rapid mapping and utilisation analysis of local areas. Equally, there are, as we noted in Section 2, applications in which humans *are* the environmental hazards. Semiconductor manufacture will continue to benefit from increased automation of chip making and micromachining, particularly if techniques such as Oxford Instrument's *Helios* x-ray lithography system becomes widely adopted. Finally, the pharmaceutical sector is a rapidly developing market for systems that include robot handling of objects such as test-tube racks, since carefully controlled trials are key to getting to market first.

Computer-integrated surgery, and medical image processing, are growing very rapidly and seem set to do so for some considerable time. The drivers are the need for cost reduction and the possibility of carrying out increasingly delicate and difficult procedures demanding long, concentration-sapping operations, dexterity that relies on precisions increasingly hard to attain by humans, and sensory information that is close to the limits of perception. Good surgeons are distinguished more by their per-operative decision making than by their dexterity, and it is the combination of robot dexterity, accuracy, and resistance to fatigue with the surgeon's ability to think, plan, and dynamically re-plan an operation that is under development. Early successes have included neurosurgery, hip replacement, computer-assisted eye surgery and dentistry, and radiotherapy. Current neurosurgery almost always uses a stereotactic frame, screwed into the patient's skull, and substantial craniotomies that expose a large part of the brain surface. Developments are likely in surgical procedures that avoid the use of stereotactic frames. An area that is attracting increasing attention is minimal-invasive surgery typified by endoscopy and laporoscopy. The technical challenges are immense: from smart processing of noisy images, through optic fibre communication of those images, tool and actuator design, registration of tool position with pre-operative information, force feedback, and the design of human-computer interfaces. Issues of safety are paramount. A related area that is ripe for considerable development concerns computer aids for the sensorially deprived. We are currently working in all of these areas.

We will see massive developments in sensor-guided robot vehicles. As little as nine years ago, mobile robots essentially comprised wire-following vehicles. At that time, we wrote a proposal to the EPSRC to extend the GEC Electrical Products robot to add obstacle detection and avoidance, and offered the opinion that robot vehicles might find other application areas within ten years. The project was funded (and was successful) but the referees dismissed the idea of non-manufacturing applications as ivory tower nonsense. Nowadays, robot vehicles have already been fielded in: civil construction for pile driving, agricultural vehicles (for example for selective crop spraying), for mowing lawns and similar regions, in mail sorting offices, and for making deliveries around hospitals and large office complexes. The Electrolux company has described in a brochure a prototype robot vaccuum cleaner that can reach under chairs and beds, and avoids obstacles such as children! Each of these applications has, if one may be excused the pun, a great deal of mileage left and it is sure that robot vehicles will develop rapidly over the next few years. The technical challenges are largely to do with sensor-guided control, with trajectory planning, and with navigation through narrow clearances and, as the inertia of the system rises, amongst moving obstacles.

Closely related to developments in robot vehicles are those in agriculture. Prototype systems have already been constructed for selective crop spraying. There have also been prototype systems constructed for harvesting, though to date they are very crude. One can imagine the potential market for a grape picking robot, though that probably reflects the fact that I have spent a lot of time recently in France, and that I enjoy wine! Whether French students and other groups of workers traditionally involved in harvesting would be enthusiastic is a different matter. Robots are also increasingly able to make compliant moves: follow a trajectory, that is try to adhere to a set of position constraints, while respecting force constraints. An example of compliant trajectory following in agriculture is the sheep shearing robot. There are plenty of others, from grooming to butchery.

Traditionally, one of the major justifications for incorporating robots in a production system has been their flexibility, born of reprogrammability. This encourages a set of dreams that have been flirted with but which remain largely unrealised. That does not diminish their potential! Key issues concern the automatic design of workspaces, particularly for parts that are large and difficult to handle. In essence one attempts to design the workspace to enable construction using robots. One project underway in our Laboratory at Oxford involves designing programmable fixtures for the manufacture of the wings of Airbus aeroplanes. More generally, and consistent with the aims of the EPSRC *Integrated Manufacturing Programme*, one might hope that the developing capabilities of robots could be taken into account in the design of business processes, including such issues as build to print and closing the loop between design and manufacture. Old themes that are likely to recur.

So far, the future seems to be mostly an extrapolation of the past, and that seems about right, for the next decade or so at least. The remaining applications in this section are far more speculative.

The biggest potential development in robot applications is in the home. Robots in the home? Well, who would have thought, in the mid-1970s, that twenty years later a majority of houses in the UK would contain several computers and at least one laser! Consider the kitchen: "smart" microwave ovens may soon be followed by more dextrous food processors, and equally smart ovens that not only control temperature, but can also carry out whatever actions are needed, such as basting, shaking, or turning the joint over. We already noted the Electrolux autonomous vaccuum cleaner: there are lots of other unpleasant jobs! My favourite would be a robot to mow the lawn, a relatively easy task except that my wife is quite protective of her plants! As is always the case in robotics, the demonstration case is easy: a lawn with clearly defined edges, a uniform green colour and even texture, no apples or deckchairs lying about, and kindly lighting. Relax any of these constraints and the system needs rapidly to get considerably smarter. Second, for me, comes cleaning the windows, a paradigm example of compliant trajectory following. Third comes ironing, but that needs to wait for the final paragraph in this Section! I may live to regret the following sentence, but I think that robot games and toys represent a potential market that dwarfs *all* the others in this Section. Video games have developed into an amazingly large market very rapidly; but how much more interestingly it would be if the creatures in the game could be reified and not just live inside a computer, if they could become animated dolls. I believe that the work that Professor Inaba has been doing represents a first, tentative step in the direction of real toys that one can interact with.

Robots are supposed to enable customisation. Wouldn't it be wonderful if they were to be used in that way in retail, in essence to encourage the return of the bespoke shoemaker and tailor! Like most people's, my feet are slightly different sizes, so I have the choice between one shoe that fits perfectly and one that is slightly loose or tight, or pay for custom-made shoes. Since I am quite hard on shoes, I never choose the latter course; but it doesn't stop me wishing I could. Can it really be so hard to design shoes just for me, starting from a CT scan of my feet? I wouldn't insist on the shoes being made while I wait, though that would be nice. The same goes for buying trousers and suits. Am I really the only person that avoids buying suits because it makes me feel like a freak? Systems that compute, quite rapidly, a good approximation to the three-dimensional shape of a person have been developed, so the technical question becomes whether one can plan and then manipulate flexible materials. Good pioneering work was carried out at Hull University on this problem. The same technology is essentially required to automate ironing. The market opportunity in each of these cases is the better informed, hence satisfied, customer, and the ways in which IT and robotics can contribute to that end. My colleagues Andrew Zisserman and Ian Reid are working on a Foresight Challenge project with CRL, John Lewis, and MFI that is similarly aimed at better informed customers. The aim is to generate from stereo vision a three-dimensional model of a room, replete with its walls, doors, and windows, then use the model as the basis for a virtual reality system in which curtains can be hung from the windows, opened and closed, the lighting changed, or in which furniture can be fitted. This is a first step along a very promising road.

5 A final word

There is a world of difference between the robots of science fiction and those fielded in industry. There is also a considerable difference between the robots that are being developed in research laboratories and those fielded currently in industry. Remarkably, hardly any of the cutting edge work sketched in Section 3 was funded by private industry, particularly the work underway in Japan and the USA. This does not square with current research funding policy in the UK, nor with UK perceptions of how research is funded in those countries. Nearly all of the work described has been funded by government agencies, or directly supported by university funds. This could well pose a problem for the UK's contributions to future work in the area unless there is a dramatic change in funding policy on the part of government or industry becomes less risk averse. Who will develop domestic applications of robots? Electrolux, apparently, for vaccuum cleaners; but I doubt that we will see either Toys R Us or Framework 5 sponsoring major research projects in robot dolls despite the massive potential of the market! Maybe my European colleagues and I will simply have to do it ourselves, and persuade investors of our case, much in the fashion of Brunel.

References

[1] V. Scheinman. Robot world - a multiple robot vision guided assembly system. In *Robotics Research: the 4th International Symposium*, Santa Clara, August 1988. MIT Press.

[2] Georges Giralt and Gerhard Hirzinger. *Robotics Research: the Seventh International Symposium*. Springer, 1996.

[3] R. W. Daniel and R. McAree. fundamental limits of performance for force-reflecting teleoperation. *International Journal of Robotics Research*, 19, 1997.

[4] Michael Brady. *Robotics Science*. MIT Press, 1989.

[5] Russell H. Taylor, Stéphane Lavallée, Grigore C. Burdea, and Ralph Mosges. *Computer-Integrated Surgery*. MIT Press, 1995.

[6] A. Blake and M. Taylor. Planning planar grasps of smooth contours. In *Proc. IEEE Int. Conf. Robotics and Automation*, pages 834–839, 1993.

[7] Andrew Blake and Alan Yuille. *Active Vision*. MIT Press, 1992.

[8] Jocelyne Troccaz, Eric Grimson, and Ralph Mosges. *CVRMed-MRCAS'97*. Springer Lecture Notes 1205, 1997.

MATERIALS

mart materials and structures

S M BISHOP BSc, PhD
ctorate of Corporate Research, Ministry of Defence, London, UK

A smart structure senses its environment and reacts beneficially. The concept is biomimetic, ie, it is borrowed from nature, and in recent years it has been the subject of much new innovative research. Investigations have been addressing the development of new sensor and actuator material technologies and their application in structures. The technology has potential in bridges, buildings, pipelines, space structures and transport, for vibration control, monitoring and controlling loads, damage detection and alleviation, and shape change (applicable in reflectors, aerofoils and hydrofoils). The benefits would include increased safety, increased comfort, improvements in performance and reduced maintenance costs.

1. INTRODUCTION

Nature has devised extremely clever mechanisms for sensing and responsive behaviour. Many are based on very complex chemistry, but often simpler physical and mechanical concepts are in evidence. In principle, there is scope to borrow many design ideas from nature. Such an approach is termed biomimetic and the smart structures concept, which, in recent years, has been the subject of much new innovative research, falls into that category.

A smart structure is one which senses its environment and reacts beneficially. Figure 1 illustrates the smart structures concept. Sensors and actuators are incorporated within the structure, which are linked to a computer system. The sensors (the nerves) sense the condition of the structure, for example, strain or another physical property. The information passes to the computer (the brain), where it is interpreted, and where the neccessary corrective action is determined. The computer then activates and controls the actuators (the muscles) in the structure to produce the neccessary response.

The technology has potential for vibration control, monitoring and controlling loads, damage detection and alleviation, and shape change. The benefits include increased safety, increased comfort, improvements in performance, increased life and reduced maintenance costs. There are a multitude of applications. Vibration control can be applied to transport and in space, and where large machinary operate. Structural monitoring, damage detection and alleviation could be applied to bridge constructions, to buildings particularly in earthquake regions, to pipelines in the off-shore industry, and in aircraft. Shape control finds application in unloaded and lightly-loaded structure such as reflectors and collectors of radiation for communications, astronomy and military applications. In more highly-loaded structures such as propellers and aerofoils, shape change could be used to control hydrodynamic and aerodynamic performance. Again this is biomimetic; birds and fishes change their profile to modify their movements through air and water.

In low-load structures, the technology may be able to be achieved relatively simply by developing existing technology, but smart-structures technology in more highly-loaded structure presents a challenge and is a real 'vision for the future' that requires considerable research and development. This paper addresses these more demanding applications.

2. SENSOR AND ACTUATOR MATERIALS AND DEVICES

There are a number of sensor technologies which could be used in smart structures. The sensors could be traditional strain gauges. Piezoceramics and piezopolymers could be used to measure displacements. Micromachined silicon sensors would be particularly attractive as strain sensors because the electric circuitry could be etched onto the silicon and the devices could be interrogated remotely thus reducing weight with a wireless system.

Sensing technologies using various types of optical fibres are being developed for both damage detection and strain measurement; strain monitoring utilising optical fibres with Bragg gratings is well developed and is already being demonstrated in some applications. In general, separation of strain and temperature effects is an important issue. Damage detection generally relies on some deformation being experienced by the optical fibre which perturbs the signal.

Electroceramics can perform both sensing and actuation roles in the same system using feedback control. Lead zirconate titanate (PZT) is well characterised and has been demonstrated for vibration control and realignment in lightly-loaded applications. Shape memory alloys can also be used as actuator materials and can impart higher forces than PZT. In general however, currently available actuator materials impart forces that are too low for use in highly-loaded structures, and the search is on for higher force, higher strain actuators. In particular phase-change ceramics show promise.

3. EMBEDDING TECHNOLOGY

Sensors and actuators can be surface-mounted on the structure. This allows existing materials and structural technology to be used and facilitates repair of the smart system. However this arrangement, with the sensors, actuators and wiring exposed, could be susceptible to accidental damage. This may also interfere with the smoothness of aerofoil and hydrofoil surfaces. A more robust and neater solution is to embed sensors and actuators in the material, which has the advantage that they can be placed at any position through the thickness. Fibre-reinforced plastics are natural contenders for smart structures because the smart technology can be incorporated relatively easily during the manufacturing process. The embedded devices would have to survive cure of the composite and still perform their intended function, and the performance of the composite material would have to be maintained.

The ideal smart technologies are fibre-like since they can be laid-up parallel to the reinforcing fibres without distorting the fibre alignment; optical fibres, shape memory alloy wires and conducting wires all fall into this category. They must be thinner than the prepreg layer to prevent distortion of the fibres in adjacent layers which could initiate fibre microbuckling. Looking ahead into the future, fibre-like smart materials might be introduced into the prepreg during its commercial manufacture.

Figure 2a shows some results for a multidirectional carbon-fibre composite containing various embedded materials. Providing the geometry was chosen sensibly, the compressive strength was relatively unaffected. In Figure 2b, the compressive failure strain decreased as the area of embedded strain gauges increased. The fatigue life of embedded gauges also decreased with size but small gauges would survive the life of a structure under working strains.

4. SMART STRUCTURES RESEARCH

During the last five years, research has been carried out at the Defence Evaluation and Research Agency (DERA), Farnborough, on smart aircraft structure. Much work has been focussed on assessing the potential of different sensor and actuator technologies. The more advanced research, aimed at their demonstration for application in structures, is described.

A cantilevered wingbox structure has been constructed to allow smart technologies in a replaceable skin to be evaluated under realistic aircraft loads in conjunction with instrumented impact of the skin. To date, the set-up has been used for assessing embedded optical-fibre sensing. Distributed sensing using birefringent fibres has been developed (1). In Figure 3, a typical output shows an increase in one component of polarization due to the presence of impact damage; the time delay gives the damage location.

An adaptive carbon-fibre composite demonstrator, in the form of a cantilevered beam, has also been developed which makes use of embedded shape-memory alloy wires as actuatorr materials. The demonstrator is shown schematically in Figure 4 with surface-mounted strain gauges for monitoring strain, embedded thermocouples for monitoring temperature, and embedded shape-memory alloy wires off-set from the neutral axis to produce bending forces. The insulated wires had to be prestrained in tension and clamped at an elevated temperature, to produce a martenistic cystalline structure at room temperature. Actuation was controlled by the resistive heating effect of an electric current through the wires, which caused contraction of the wires and a resultant compressive force as they returned to their austenitic form. Load control was demonstrated by placing a range of weights at the end of the beam as in Figure 4, and then countering the bending strains using the shape-memory alloy wires; it can be seen in Figure 5a that this was successfully achieved although the response was not very fast. Shape control was also demonstrated by actuating the system to hold a controlled strain over a period of time (Figure 5b). Strain and temperature were controlled using feedback.

The potential for buckling suppression, using various actuator materials, was modelled in a curved carbon-fibre composite structure containing a hole (representing gross damage). The 4mm thick panel (Figure 6) was designed so that, with the hole, the buckling strength fell below design-limit load. The aim was to restore the panel's performance to an operable level by redirecting the load path and reducing the stress concentrations at the hole. Electroceramic actuators positioned in a grid were selectively actuated in tension, compression, or bending, to counter stresses due to the applied load. Insufficient force was produced by PZT but phase-change ceramics appeared to have the capability.

5. SPECULATIVE RESEARCH FOR THE FUTURE

A very attractive development for smart structures would be structural materials with self-adaptive capability and no parasitic weight. This is a very speculative 'vision for the future' but new research within DERA is examining the potential for developing such materials using the biomimetic approach. The variable-stiffness mechanisms found in mutable collagen have been studied in sea cucumbers, in collaboration with the University of Reading. At DERA the synthetic possibilities are being investigated (2). Mutable collagen is essentially a short-fibre composite and the variable-stiffness phenomenon appears to occur as a result of variable load transfer into the reinforcing fibres. Load transfer occurs as illustrated in Figure 7a. It is suggested that, in sea urchins and sea cucumbers, the modulus of the polysaccharide matrix is changed by a reversible chemical effect, and thus the load transfer length and the reinforcing ability of the collagen fibres is changed. Changes in composite modulus have been calculated using the Cox model, using representative matrix properties and fibre dimensions for sea urchin (Figure 7b), and for a synthetic possibility, a short fibre composite (Figure 7c). It can be seen that a reasonable change in composite modulus from 70 GPa to 90 GPa could be obtained for carbon fibres 90 μm in length if the matrix modulus could be triggered to change from about 1 to 3 GPa.

Some preliminary modelling was carried out to determine if this composite stiffness change could be used in the skin of a wingbox to produce useful adaptive behaviour. The tapered skins were given a (0°, 90°, -45°, +45°) lay-up (Figure 8) and the twist induced by synthetic variable-stiffness material in 45° layers was evaluated under load; the effect was rather like aeroelastic tailoring and useful twist was induced. Variable stiffness materials would be much lighter than embedded lead-based electroceramics and weight savings could be significant.

ACKNOWLEDGEMENTS

The research reported here was carried out in the Structural Materials Centre, DERA, Farnborough, as part of Technology Group 04 of the MOD Corporate Research Programme. The author, who led the team in her previous post, would like to acknowledge the scientific contributions of C J Doran (embedding technology, shape-memory alloys, variable-stiffness material), I P Giles of IPG Associates and P A Lloyd (optical-fibre sensing), S Singh and P J Hopgood (buckling suppression, variable-stiffness structure).

REFERENCES

1	P Lloyd C Doran	'Military applications of smart materials and structures' Journal of Defence Science. 1, No 3, 338 - 351 (1996)
2	C J Doran S M Bishop S Singh	'Use of biomimetic stiffness-change mechanisms within aircraft structures', Proceedings of the SPIE Fourth Annual Symposium on Smart Materials and Structures, San Diego, USA, 3040-47 (1997)

C527/

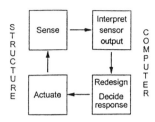

Figure 1 The smart structures concept

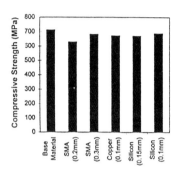

a) Compressive Strength

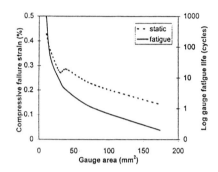

b) Embedded strain gauge

Figure 2 Properties for embedded materials in a multidirectional carbon-fibre composite.

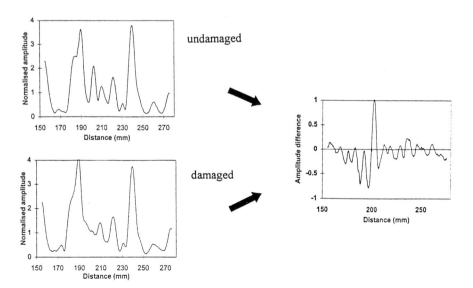

Figure 3 The use of birefringent optical fibres to detect and locate impact damage.

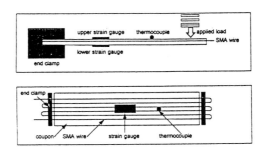

Figure 4 Cantilever beam containing shape-memory alloy wires.

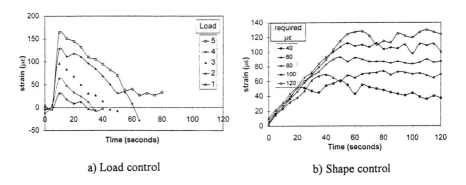

a) Load control   b) Shape control

Figure 5 The use of shape-memory alloy wires to control load and shape

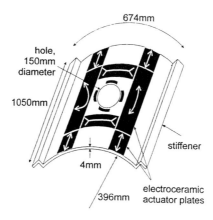

Figure 6 The use of smart actuators to suppress buckling within a curved panel

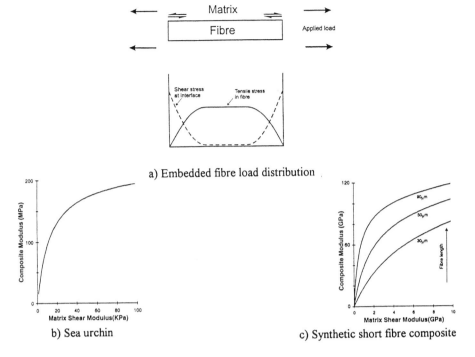

a) Embedded fibre load distribution

b) Sea urchin

c) Synthetic short fibre composite

Figure 7 The concept of variable stiffness materials based on biomimetic concepts

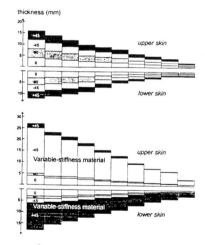

Figure 8 Skin structure of:
(a) a conventional wing-box (quasi-isotropic skins)
(b) smart adaptive wing-box with skins containing stiffness-changing material.

Designing new materials by computer simulation

D R ADAMS BSc, PhD, MMA
Director, Mechanical Engineering and Metallurgy, T&N Technology Limited, Rugby, UK

SYNOPSIS

Intense competition in industrial markets today demands that new products are developed in as short a time as possible. Materials development forms an integral part of the product development process and it is therefore essential that techniques are available to shorten the time taken to design and optimise new materials. Traditionally the materials development process has been largely empirical. However new techniques based on computer simulation are becoming available to assist the materials engineer in his task. This paper examines the application of such techniques and considers the potential impact of emerging technology.

1. INTRODUCTION

The development of a new material typically depends to a significant degree on empirical methods. This is due to an incomplete understanding of the relationships between properties and factors such as composition, physical and chemical interactions between the constituents and the processing route. Whilst capable of realising improved materials it is unlikely that continued reliance on empirical methods will achieve a fully optimised composition in an acceptable development time. The time to market for new products is a critical aspect of competitiveness and increasingly there is a need to seek more rapid development routes.

This paper explores the impact of current and emerging computer aided methods for materials development. The majority of the examples used in this paper will be drawn from the author's experience in the development of components for automotive applications.

2. THE MATERIALS DEVELOPMENT PROCESS

A materials development programme usually begins with a description of the function of the material and target improvement. Function is usually expressed in terms of some aspect of component performance, for example in the case of a gasket it might be the ability to seal at high temperatures, whilst for a piston it might be durability expressed in terms of engine life. The driving force for improvement comes from some deficiency in the function of a component which is then translated into a material property or set of properties that require enhancement. At this point the materials engineer must apply his knowledge and experience to firstly conceptualise and then produce various formulations which are evaluated by testing. An iterative procedure to optimise performance will generally lead to an improved material.

There are several difficulties with this process. Firstly, the target for improvement may not be directly related to a physical property of the material. For example, the wear performance of a material cannot today be reliably predicted from any simply measured set of properties. making it necessary to resort to simulated environmental testing in almost every case to determine whether the desired improvement has been achieved. Secondly, the processing conditions will interfere with the kinetics of the physical and chemical processes at work during the manufacture of the material, again necessitating experimental work. Thirdly, the environment of the component is often ill-defined so that whilst it may be desired to improve, say, fatigue life, the conditions of stress within the material may only be approximately known thus making it difficult to define the property improvement needed to achieve a desired increase in performance.

The main consequence of these difficulties is that materials development is relatively slow and not always successful. Computer techniques are now emerging that address these difficulties. The remainder of the paper is devoted to a description of these techniques with illustrative examples drawn from the work at T&N.

3. COMPUTER AIDED MATERIALS MODELLING

A range of modelling techniques are being used within T&N to increase the speed and accuracy of materials development. A brief overview of these techniques is given below.

3.1 Designed Experiments

One of the earliest methods for materials optimisation is the designed experiment. Originally developed in the 1920's for application in the agricultural industry the technique has evolved considerably during the last 20 - 30 years, particularly as a result of work by the Japanese, into what has become known as the Taguchi method. This well known technique enables the influence of a set of variables to be determined using an economical matrix of tests (partial factorial) thus avoiding the need to evaluate every permutation of the factors (full factorial). With the advent of PC's and user-friendly computer programs it has been possible to simplify the use of this technique to the point where it is now accessible on a routine basis.

The use of designed experiments essentially provides a systematic framework to determine parameter effects and their interactions. Having determined these effects it is a relatively simply step to use the resulting relationships to 'predict' an optimum composition. An example of a typical 3D response determined by the software is shown in Fig 1.

The shortcoming of the designed experiment is that the optimisation process is entirely empirical. It simplifies the problem inasmuch as it does not require any real knowledge of the environment of the material. Thus although it is an effective development tool it rarely provides any real insight into the problem and its solution.

3.2 Numerical Analysis

The rapid development of finite element analysis (FEA) and associated computer hardware has enabled huge strides to be made in the application of numerical analysis in materials development. The means are now available to calculate field variables such as temperatures, stresses and distortions in almost any component, given a reasonable knowledge of the material properties and boundary conditions (Fig 2). Such simulations have also been extended to manufacturing processes such as metal forming and casting (1, 2)

Numerical analysis assists materials development by characterising the environment of a component so enabling the operating temperatures, stresses, etc to be calculated. This can form the basis for materials selection and/or indicate the desired target properties of a new material to meet the demands of a given application. In short numerical analysis provides a quantifiable development target. However, as with the designed experiment the actual process of developing an improved material still relies entirely on the knowledge and experience of the materials engineer. In order to complement the insight that numerical analysis provides there is a need for a means of predicting the properties of a material directly from its composition. Thermodynamic modelling offers this promise.

3.3   Thermodynamic Modelling

Thermodynamic modelling describes techniques which attempt to simulate the physics and chemistry of metal alloy systems from basic thermodynamic principles. The technology is still relatively in it's infancy, probably being similar to where finite element analysis was about 15 years ago. Several software packages exist to calculate the phases present in alloy systems. The essential principle of this and similar software is that mathematical models incorporated in it allow equilibria in multi-component systems to be calculated on the basis of thermodynamic data from simpler sub-systems. It is thus possible to explore alloy compositions and temperature conditions for which no direct experimental data exist.

3.3.1 Microstructure Modelling

A natural development of thermodynamic modelling is to incorporate kinetic effects to allow the prediction of microstructure from the alloy composition and processing conditions. Such a tool appears within reach in the near future and will be followed by the inclusion of mechanical properties in the model, giving a complete alloy development tool. T&N is working with several UK universities with a view to applying this tool to materials for pistons, bearings and cast iron camshafts.

4. EXAMPLES

4.1 Materials Optimisation by Design of Experiment

Experimental design has been used extensively to optimise formulations for gaskets, friction products and polymers. As an example the optimisation of a brake pad formulation for use as a competitions material for cars and motor cycles is described.
The starting formulation was complex and contained eight ingredients, making

experimentation complex. It was required to optimise several responses such as friction coefficient up to temperatures of 700 °C and pad and disc wear, though other factors such as speed sensitivity, stop shape and manufacturability needed to be considered. An experimental design was derived which enabled all of the important factors, together with a number of material interactions, to be optimised. The experimental design stipulated the manufacture and test of 12 formulations. From the test results a mathematical model was constructed which showed how each ingredient affected the properties of interest. Figure 3 which shows the 3D response surface of 3 ingredients of the pad with respect to friction and wear. Using this model the formulation was optimised for maximum friction coefficient, minimum disc and pad wear and speed sensitivity. The optimised formulation was manufactured and its performance found to be close to the prediction.

The developed material was successfully launched commercially where it is used in the competitions market for Formula Vauxhall and a number of the larger motor cycles such as 600cc and 750cc superbikes.

4.2 Numerical Analysis

There are numerous examples where the use of FEA and associated numerical techniques have been successful in aiding materials development. The first example is taken from an analysis of the interaction of an engine valve and the insert against which it rests when closed. One of the trends in engine development is towards higher combustion temperatures which are beneficial to fuel economy and engine emissions. However hotter engine running is detrimental to components unless their temperatures can be reduced. The engine valve is particularly vulnerable as it operates at temperatures of over 500 °C. Much of the heat energy into the valve is conducted away through the valve seat. The thermal conductivity of the insert therefore has a significant influence on the heat flow and hence valve temperatures. The aim of the study therefore was to examine the effect of insert conductivity on component temperatures. An FE model of the region in the cylinder head was used to predict operating temperatures (Fig 4) and hence calculate the effect of valve seat insert conductivity on valve and valve seat insert temperatures. The results (Fig 5) show clearly that a high valve seat insert conductivity is beneficial in reducing the temperatures of both the seat and valve. Since many valve seat inserts are manufactured by powder metallurgy, the process provides a convenient means of achieving the desired properties by infiltrating the insert with copper during sintering. The process allows the control of thermal conductivity to meet the demands of the application.

A second example relates to materials used for disc brakes pads. One of the undesirable phenomena in a disc brake is for hot spots to be generated during braking which are responsible for cracking of the brake rotor. The cause of hot spotting is complex but is related to the tendency for the pressure distribution between the pad and disc to drift to a larger radius and become more concentrated - Fig 6. Results from vehicle applications have shown that a reduction in the modulus of the pad can have a beneficial effect on the maximum temperature by creating a more even pressure distribution. The predictions of the model, shown in Fig 7, demonstrate very clearly the beneficial influence of a lower modulus. A further use of this model demonstrated that increasing the wear rate of the friction material was also an effective means of creating a more even pressure distribution, Fig 8. This resulted in a 70 °C reduction in predicted maximum temperature. Although an increased wear rate would be detrimental to wear life materials design is always a compromise between competing factors depending on the particular application.

The preceding examples showed how modelling can be used to identify targets for materials development. Another, complementary use of modelling, considers the effect of changes in component geometry to extend the application of an existing material. All too often component failure is attributed to a material problem when the real solution lies in a flaw in the basic geometry. This is illustrated in the following example which concerns the design of an exhaust manifold gasket for a reciprocating engine. The gasket was found to leak during engine testing. Analysis of the exhaust manifold and the flange seal showed an area of low sealing pressure at a point which matched the location of observed gas leakage (Fig 9). The reason for the low sealing pressure can be seen in Fig 10 to be due to the very uneven flange deflections. One of the proposed design solutions was to stiffen the flange by adding a web, Fig 11. This resulted in a 12% increase in the minimum sealing pressure which was deemed sufficient to seal the flange. This type of problem is probably more common than is realised. In most cases more time spent on analysis at the conceptual stage would avoid such problems.

The condition of a material can have a profound effect on its performance. For example the existence of surface defects or adverse residual stresses can seriously reduce the design life of a component. This is not necessarily a problem with the material. More often than not the problem lies with the manufacturing process. The last example shows how modelling can provide insight into these types of problems.

Aluminium pistons are usually manufactured by a casting process. Sources of surface defects in the piston include interruptions in the flow of metal during casting, porosity and contaminants. Although such defects are not common when they do occur the possible risk of a crack propagating from the surface has to be considered. This can be achieved using the theory of fracture mechanics. Fracture mechanics defines the relationship between the stress at the surface that will cause a crack to grow from a defect of a particular shape and size. A typical relationship for an aluminium alloy is shown in Fig 12. By incorporating this relationship into the FE analysis of, say, a piston it is possible to produce a plot, Fig 13, identifying those regions which are sensitive to defects of a particular size. Conventional stress analysis may indicate that such areas are safe yet the existence of a defect increases the risk of premature failure in service. The result of such an analysis might be to change the manufacturing process to eliminate the defect. Alternatively showing that the defect will not propagate avoids scrapping what is essentially a sound component.

4.3 Thermodynamic Modelling

To date thermodynamic and microstructural modelling are still at an early stage in their development. Exploitation is constrained by the need to generate data for a wide range of alloy systems. Through a number of university programs T&N is focusing on aluminium silicon alloys and cast irons. Promising correlations between measured and predicted property data are being achieved (Fig 14 & Table 1) and it is now routine practise to predict specific heats, critical temperatures, such as solidus and liquidus, and the composition and amount of phases at any temperature. Such data are already being used to guide materials development programmes within T&N. For example by systematically varying the percentage of a given element it is possible to examine the occurrence and stability of certain phases which may be beneficial or deleterious to properties. Other examples include the determination of melting and pouring temperatures for use in casting operations and for quality control of raw materials. They can also be used directly in solidification modelling software to support component and process design without even making the alloy in question.

5. THE FUTURE

With the exception of the aerospace, nuclear and civil engineering industries, up to about 15 years ago FEA and other computer tools were still relatively new. As a consequence the materials engineer had access to little information to help define materials development targets. This placed a heavy emphasis on component testing to confirm any improvement in properties. During the intervening period FE techniques and computer hardware have made rapid strides giving the designer an opportunity to achieve significant insight into the nature of the materials application. Whilst this has increased confidence in targets for materials development there is still reliance on what is essentially a process of trial and error to achieve the targets. As microstructural modelling develops however it is anticipated that it will provide the means of reducing much of the emphasis on trial and error. By this means it will be possible to predict the properties of an alloy from a knowledge of only its composition and kinetic information.

The Holy Grail for materials development would be an ability to link the component model with the material model to predict the required composition directly from the application requirements. One possible tool for achieving this is neural network modelling. Neural networks are designed to imitate the ability of the human brain to make sense of complex and incomplete data. In addition neural networks can model processes with many variables without a detailed knowledge of the complex interactions in the system, thus avoiding the need to work from first principles. Although still limited in its application materials problems successfully tackled in this way include the modelling of weld metal toughness (2) and physical property prediction in nickel based superalloys (3). Again there remains the constraint of the volume of data required to enable a neural network to produce reliable predictions. A technique which may offer some help uses so-called fuzzy algorithms which have an ability to deal with uncertainty and incomplete data sets. Such an approach will reduce the amount of data needed in the future.

CONCLUSION

Computer simulation is having a dramatic impact on the materials development process. Future developments in this field are likely to revolutionise the material engineers task of designing new materials and significantly reduce the time taken to achieve optimised properties.

REFERENCES

1. Towers, O.L., Process Modelling, Seminar on Numerical Techniques, Institute of Metals, London, (1989)
2. MacKay, D.J., Bayesian Non-Linear Modelling with Neural Networks, University of Cambridge Programme for Industry, (1995)
3. Jones, J., Neural Network Modelling of the Tensile Properties of Nickel-Base Superalloys, Dissertation submitted for the Certificate of Postgraduate Study at Cambridge University, (1994)

Alloy	ΔH_{fusion} (Jkg⁻¹)		Solidus Temp (°C)		Liquidus Temp (°C)	
	Measured	Predicted	Measured	Predicted	Measured	Predicted
A	531	600	539	532	618	614
B	501	515	542	541	573	596
C			1297	1289	1328	1335
D			1310	1318	1380	1374
E			1292	1357	1375	1416

Table 1 : Comparison of Measured and Predicted Property Data using Thermodynamic Modelling

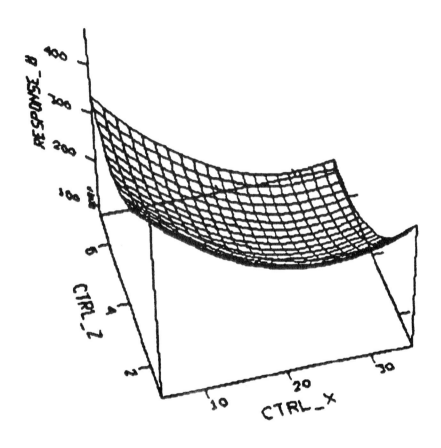

**Figure 1 : Typical 3-D Response Surface
Derived from a Designed Experiment**

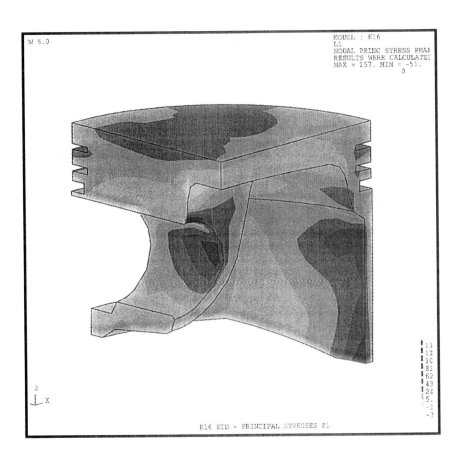

**Figure 2 : Typical Output from a Finite Element
Analysis of an Automotive Piston**

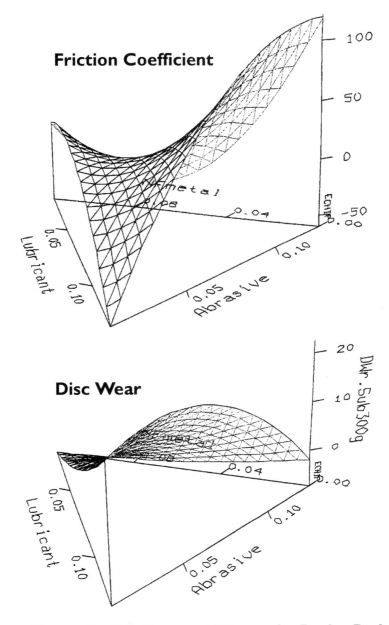

**Figure 3 : Friction and Wear of a Brake Pad
as a function of its Composition**

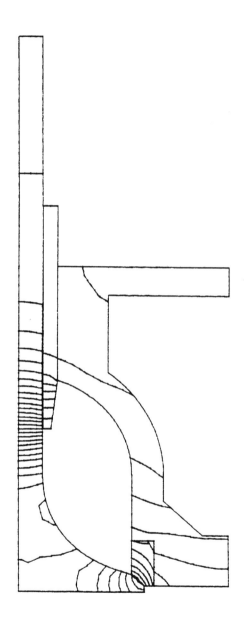

**Figure 4 : Predicted Temperatures in a Valve,
Valve Seat Insert and Valve Guide**

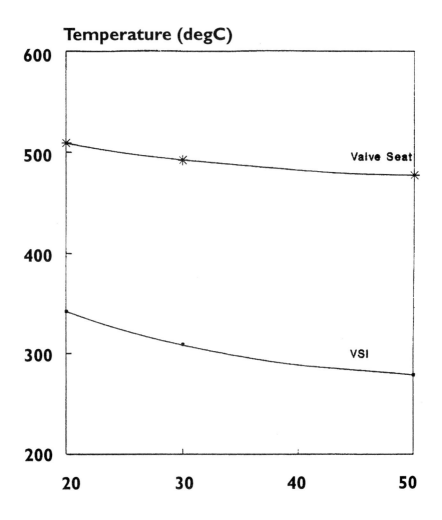

Figure 5 : Effect of Valve Seat Insert Thermal Conductivity on Maximum Insert and Valve Seat Temperatures

C527/(

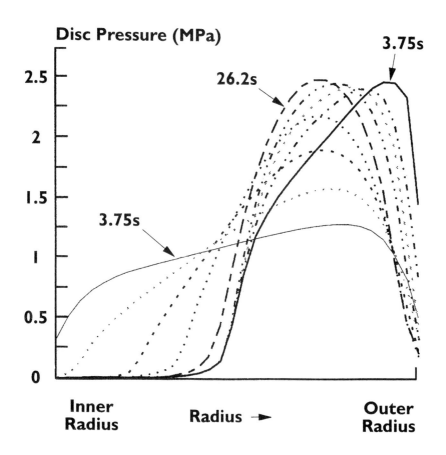

**Figure 6 : Localisation of the Pad-Rotor Pressure
Distribution During Braking**

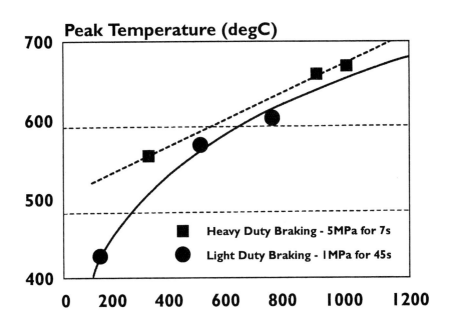

Figure 7 : The Effect of Disc Pad Modules on Maximum Rotor Temperatures

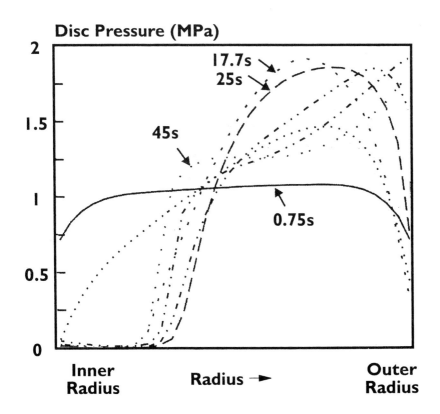

**Figure 8 : Localisation of Pad-Rotor Pressures
with Increased Wear Rate**

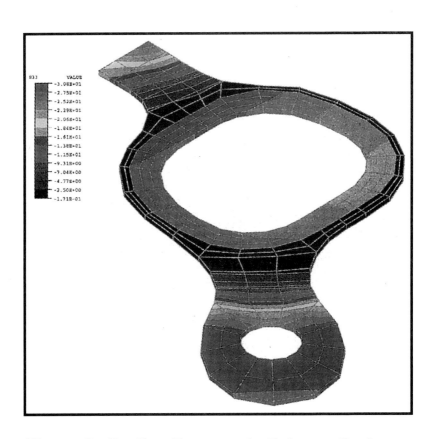

Figure 9 : Sealing Pressure in Exhaust Gasket

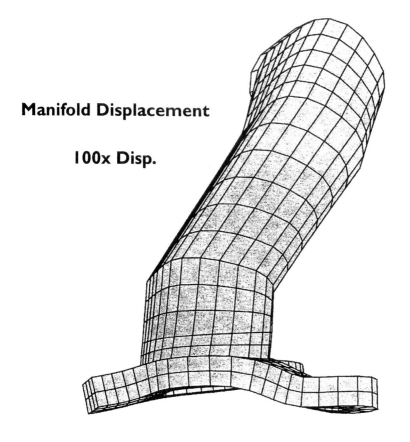

Manifold Displacement

100x Disp.

Figure 10 : Deflections in Exhaust Flanges

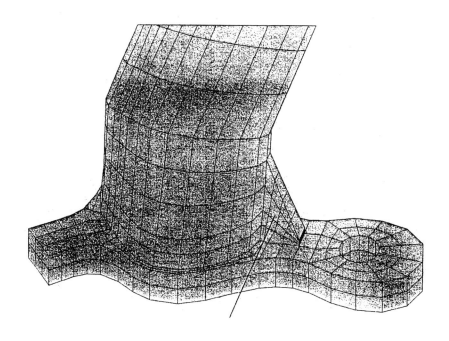

Figure 11 : Stiffening Web in Exhaust Flange

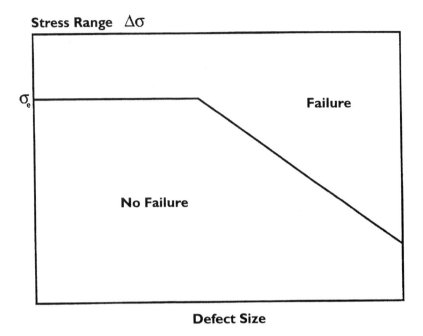

Figure 12 : Typical Relationshhip between Stres Range and Critical Defect Size for an Aluminium Alloy

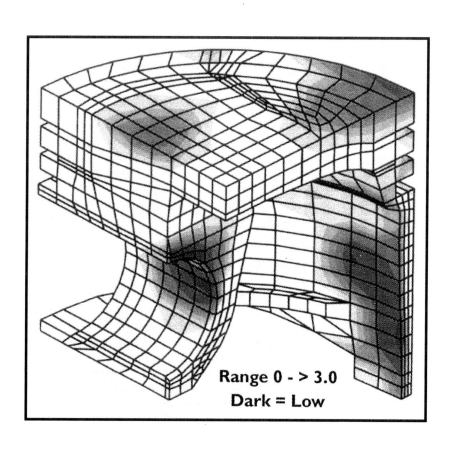

Range 0 - > 3.0
Dark = Low

Figure 13 : Maximum Permissible Defect Size in an Automotive Piston

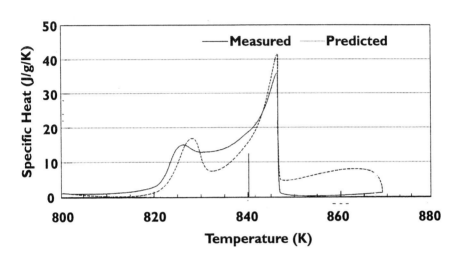

Figure 14 : Comparison of Measured and Predicted Specific Heat for an Aluminium Piston

dvanced composites

EAD FEng
naging Director, G Maunsell & Partners Limited, Beckenham, UK

1.   IN THE BEGINNING

Advanced composites have evolved in natural forms and are all around us. They are nature's solution to providing strong, impact resistant, lightweight structures. In bones and ligament they provide the perfect structures for fast moving creatures, having just the right balance between weight, strength and durability. In trees the fibrous structures flex with the weight of foliage and are able to bend in the wind without breaking.

For early man these materials and stone were the only engineering materials and for many thousands of years they formed the essential structural elements of life. The rapid development of metals in the Industrial Revolution meant that when the Institution of Mechanical Engineers formed in 1847 it was the beginning of a unique period of 150 years when metals dominated engineering design.   'Materials' become synonymous with 'metallurgy' and the world's universities taught courses and awarded degrees in metallurgy and metal science. Graduates had barely heard of polymers and composites and were unaware of their roots in nature.

However in the last twenty years that has changed. The steel industry worldwide has stopped expanding and the rate of development of other metals and alloys have slowed. The production of high strength polymers, ceramics and structural composites is expanding. We are now in the middle of another revolution, a transition from the steel age to one dependent or other more advanced materials.

It is interesting to note that the birth of the commercial plastics and composites industry lay in the invention of 'Parkestine' by Alexander Parkes who received an award at the Great Exhibition in 1862, not long after the birth of the Institution. Developments really took off much later with the pressures created on engineers in the last war to achieve better aircraft performance. This is summarised in Table 1.

There is now an increasing demand for strong, lightweight materials which can be tuned to particular needs and are durable. Public acceptance is developing through a reawakening of that primal understanding that fibrous composites are excellent structural forms, and can do anything and more that metals have done.

The materials are everywhere already and becoming more visible every day. Mrs. Jones's quality of life depends on these materials. Who is Mrs. Jones?

Table 1 - Brief Historical Highlights of the development of Reinforced Plastics

1862	-	Alexander Parkes showed `Parkestine'.
1917	-	Formica Products Company developed laminated products.
1930	-	Glass fibre research initiated by Owens Illinois and Corning Glass works.
1929	-	British Plastics Federation formed.
1933	-	C. Ellis patented unsaturated polyester resin.
1942	-	First reinforced plastic components produced in USA.
1943	-	First use of FRP in aircraft airframes in USA.
1945	-	First filament winding of RP Hoops at Bassens Industries.
1947	-	First GRP washing machine.
1951	-	First GRP boat hull built by W&J Todd. UK.
1953	-	First GRP body panels for cars - Corvette. USA.
1955	-	First FRP gun barrels.
1956	-	First filament wound rocket motor.
1957	-	First Lotus Elite car produced using GRP. UK.
1958	-	Graphite fibres produced for first time.
1961	-	Continuous graphite filament commercially produced.
1962	-	Vinyl ester resins produced by Dow and Shell.
1963	-	GRP petrol tanks first produced.
1968	-	First SMC bumpers mass produced - Renault 5.
1971	-	Kevlar aramid fibre available from Dupont.
1980	-	First hand lay-up GRP Road Bridge - China.
1984	-	First FRP complete airframe - Avtek 400.
1986	-	First Carbon fibre aircraft Beech Starship.
1988	-	First advanced composite bridge enclosure - A19 Tees Viaduct. UK.
1992	-	First fully bonded advanced composite cable stayed bridge - Aberfeldy UK.
1992	-	First FRP 'Plastic' car engine - USA.
1994	-	First FRP firewater caissons on offshore rig. UK.
1994	-	First advanced composite road bridge - Bonds Mill UK.
1995	-	First FRP Fire and Blast walls on offshore rigs.

2. QUALITY OF LIFE TODAY

Mrs. Jones lives in Gwent with her family in a converted farmhouse. The family have a composite dinghy which they sail from the local sailing club on the Severn Estuary at the weekend. (The development of GRP mass produced boats has made boating accessible to the general public). This weekend however she is off to a residential conference in Portugal and is loading her luggage and golf clubs into the car to set off to Bristol Parkway Station. (Her new set of golf clubs are very light to carry, with their carbon fibre/boron shafts. The irons are carbon faced, producing longer drives with less spin). Mrs. Jones has been playing much better with these new clubs and is looking forward to impressing her colleagues. She sets off in her car (the car has a carbon/glass polyester moulded front end which gives the car distinctive aesthetics, low weight and drag) which is surprisingly economical on fuel for such a sporty performance. She enjoys the rather fast drive over the elegant new Second Severn Crossing with its very modern bridges on the approaches. (The bridges have full advanced composite enclosures to reduce maintenance and traffic disruption during inspections). The journey is quick and trouble free. This means she does not have a long wait at the station for the Intercity 125 train to Reading. Mrs. Jones is pleased to see the familiar shape of the 125 coming in on time. It is not as flashy as the Eurostar but still looks good after 20 years in service. (The front end of the 125 and Eurostar are advanced composite structures which reduce weight, lower energy consumption, create good aerodynamics and have excellent impact resistance).

At Reading Mrs. Jones carries her lightweight luggage (Mrs. Jones' brief case is a glass reinforced plastic one she has had for 25 years. It has survived the rigours of international travel all that time) and golf clubs to the airport bus. It is an impressive new type of coach she has not seen before, all smooth curves and glass. (The new coach has mostly FRP bodywork and floor panels on a steel frame. The FRP reduces weight, improves aesthetics and durability. The seats have FRP moulded fames).

As she arrives at Heathrow, Mrs. Jones is thinking to herself how much more comfortable and reliable the journey has been compared with experiences 20 years ago.

She gets out her laptop computer in the airport lounge to catch up on some work (the laptop has an advanced composite board for dimensional stability, low weight and electrical conductivity characteristics).

The flight to Portugal on an A320 plane is delayed by air traffic control but she arrives at the hotel in time to meet her colleagues for dinner. (The A320 has a large number of composite components throughout the aircraft to reduce weight and operating costs).

After a very pleasant weekend in the sun, Mrs. Jones arrives home late on Monday evening with driving rain coming up the Severn Estuary on a gale force south westerly. Mr. Jones says rain is coming in through the upstairs windows again, and decides that the next day he will ring up that new company who are advertising a range of replacement windows in longlife stable FRP. (FRP pultruded window frames have low thermal conductivity and similar co-efficient of expansion to glass and brick). Mrs. Jones sleeps soundly, despite the rattling windows, dreaming of that winning round of golf.

3. DEVELOPMENTS AND SUCCESSFUL APPLICATIONS IN CONSTRUCTION - INDUSTRIES AND DISCIPLINES WORKING TOGETHER

I will now examine the driving forces behind the development of the use of advanced composites which is helping to achieve a better quality of life for Mrs. Jones. In particular I will illustrate the paper with some success stories in the construction industry in bridges.

The ending of the 'Cold War' combined with worldwide recession in the developed world in the early 1990's arrested the automatic and massive investment by Governments and private industry in the defence and aerospace industries which started during the last war. Hence private industry which has invested heavily in new materials and technologies for those industries has suddenly lost its market and is actively searching for new market opportunities. The construction industry with its huge turnover and history of lack of investment has readily become a prime target. This is most evident in the USA and Japan where large investments are now being planned by Private industry in harness with Government into High Performance Construction Materials and Systems. Proposed materials include improved concrete, higher strength steel, aluminium and advanced composites and the systems include new methods of construction with robotics.

Advanced composite fibre reinforced polymer materials are the high performance materials which are likely to have the most impact on new structural forms over the next century. Their development in aerospace, automotive and marine industries has been very successful and research and prototype projects for construction applications worldwide have demonstrated considerable potential. Likely applications include:-

- · all composite structures
- · fibre reinforced concrete
- · prestressing for concrete and masonry structures
- · reinforcement and permanent formwork for concrete
- · enclosure skins for steel and concrete structures
- · steel and composite bonded structures

Particular advantages of advanced composite materials, using large proportions of carbon, aramid or glass fibres in a polymer matrix, are greatly improved corrosion resistance and durability, low weight, ease of transportion, low thermal conductivity, non-magnetic properties and lower energy consumption during manufacture and end-use. Typical properties are given in Table 2. Also complex optimised and modular system designs can be produced without high labour costs using automated production methods such as pultrusion. It is the high strength to weight ratio of these materials that will eventually allow bridge engineers to extend spans from the 1600 metres current maximum to over 5000 metres in the future.

Table 2: Typical Mechanical Properties for GFRP, CFRP, and AFRP

Uni-Directional Advanced Composite Materials	Fibre content (% by weight)	Density (kg/m3)	Longitudinal Tensile Moduls (Cpa)	Tensile Strength (MPa)
Glass Fibre/ polyester GFRP laminate	50 - 80	1600 - 2000	20 - 55	400 - 1800
Carbon/epoxy CFRP laminate	65 - 75	1600 - 1900	120 -250	1200 2250
Aramid/epoxy AFRP laminate	60 - 70	1050-1250	40 - 125	1000 1800

Prototype bridge structures are now being built all over the world using advanced composite materials for bridge decks, supporting cables, concrete reinforcement and prestressing. Aberfeldy Footbridge in Scotland is currently the World's longest span bridge built entirely out of these materials with a main span of 63 metres. (Figure 2) Design and construction of this bridge has given new insights into design and construction technology using lightweight materials particularly with respect to achievement of aerodynamic stability. An advanced composite road lift bridge, recently completed at Bonds Mill in Stonehouse, England, is carrying full UK traffic loading with vehicles up to 38 tonnes. These prototype structures are being studied and monitored to provide design data for larger applications now on the drawing board. University students have been involved in all these developments to help create a new generation of technologists who are familiar with these materials.

The development of the glass reinforced polyester Advanced Composite Construction System (ACCS) by Maunsell in the period 1982 to 1992 enabled these milestones to be achieved. The inspiration for the use of these materials in construction came from the successful research and development carried out by the Ministry of Defence at Dunfermline in the 1970's (for HMS Wilton) which showed their true potential in load carrying situations. A parallel engineering programme for ACCS was instigated by Maunsell to carry out the necessary research and development, including material scientists, production, mechanical, and structural engineers working closely with University research departments. World leading technology was developed quickly and at relatively low cost.

The foundation for the work was the early development of a Limit State design method for the materials linked to a detailed manufacturing specification. In this respect considerable help was given by the Aerospace industry and their CRAG test methods were incorporated as the basis for all destructive testing. Designers worked closely with mechanical engineers in further developing the pultrusion process used to produce the interlocking FRP profiles in ACCS. Figure 1.

4. LOOKING TO THE FUTURE

In the search for continuing prosperity in an increasingly competitive world market, Governments are beginning to realise that an efficient and reliable infrastructure is one of the most important foundation stones for continued prosperity. It is clear from the experience of the last thirty years in Europe, Japan and particularly the USA that existing (so called proven) construction technology has not delivered the reliability needed. Maintenance causes huge traffic congestion and is beginning to impact on overall economic growth. Hence the conclusion has been reached in USA and endorsed by President Clinton that "tomorrow's infrastructure must not be built using today's technology!". However it is recognised there that the evaluation and acceptance of innovative construction products is a process which is not in place and there are many difficulties to be faced as we return to a more entrepreneurial approach.

Environmental and safety considerations are now beginning to move to the forefront of all decision making processes from planning through design and choice of materials and on to operation and maintenance. The sustainability of resources, energy consumption and pollution during all project phases, safety for users and maintenance staff and noise caused by traffic are becoming increasingly important. These views are beginning to shape changes in the form of structures and the materials used in them. Advanced composites have much to offer in all these respects.

Turning particularly to bridges, one of the greatest failings of current infrastructure design is the failure to design for change of use, maintenance and replacement of parts during the full design life of up to 120 years. All other industries design their structures around this requirement, for example cars, planes and ships. The human body is a perfect example of a complex structure which is self diagnostic, can be maintained, repaired and components can be replaced readily!

It is very likely that successful future bridge systems will include more and more of these features over the next century. A typical structure might consist of a high performance concrete or steel or steel/concrete composite structural frame covered with an advanced composite, self diagnostic SMART skin. The skin would make a structural contribution and also provide corrosion protection, environmental control, monitoring functions, aesthetics and aerodynamic shaping. Access for maintenance and inspection would also be provided by the skin and this could be removed and later replaced for major strengthening, adding extensions to the structure or for changing the external appearance. Figures 3, 4 and 5 show the actual development and evolution of this technology in bridge design from a traditional steel composite plate girder design through the new award winning enclosed bridges on the Second Severn Crossing approaches to a patented new form of high performance structure for the next century.

The structural form shown in Figure 5 is called SPACES and is being developed and promoted by a partnership of materials suppliers, manufacturers, fabricators, bridge designers and constructors. SPACES draws on many technologies developed in the offshore industry.

SPACES has been developed as a total system combining key technologies and yet none are totally new. Each has a well proven background, yet never before have they been combined together in bridge construction. The SPACES System takes us away from predominantly site operations to a factory manufacturing approach where quality and reliability can be far better achieved. A key to the economic success of the system is the development of a robotic welding system for the joints between steel tubes. This work is well underway. The System application is wide ranging from relatively short 60m spans to long span cable supported decks. For the first time designers are able to adopt a "Systems Approach" to bridge engineering not for just a small range of spans, but for the complete range. SPACES adopts a system approach thereby providing inherent reliability and geometric flexibility while allowing much greater aesthetic freedom than is normally available to bridge designers.

The key to the SPACES concept is the excellent long term performance provided by the advanced composite enclosure skin. The structural form, materials and technologies in combination provide excellent long term durability. Along with that, the bridge owners will enjoy the benefits of permanent inspection and maintenance access. The minimal maintenance programme means much lower whole life costs, while avoidance of road closures means lower traffic disruption costs.

Figure 6 shows a recently commended competition entry for a bridge crossing at Poole Harbour using the SPACES system. Advanced Composites are now bringing a whole new aesthetic to bridges and buildings as they have already done for cars, trains, planes and boats.

A further glimpse into the buildings of the future is the proposed Globorama tower in London, Figure 7 a centre for environmental awareness and global communications. Advanced composite materials will be used for much of the top of the structure.

5.   CONCLUSION

The entrepreneurial fires that were burning brightly at the turn of the last century are being fanned into life again as society worldwide faces new challenges. Overcrowding and environmental problems are driving engineers to look for more durable materials derived from sustainable resources to build our infrastructure and transport systems. Structures have to be more flexible to accommodate change of use and must be able to be maintained easily . New forms of structure are emerging and the beginning of the 21st Century will be a fascinating period of change in technology, as advanced composite materials begin to move to the forefront of design thinking. It is unlikely that the materials will replace metals, but will be successfully used in combination with these to achieve improved performance. Advanced composites are also becoming an essential part of the transportation and leisure industries worldwide. Mechanical Engineers are working closer and closer with all other disciplines of science and engineering at the heart of this technology revolution.

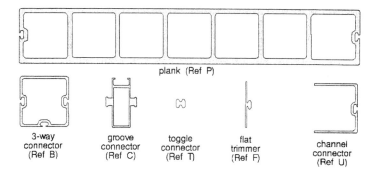

plank (Ref P)

3-way
connector
(Ref B)

groove
connector
(Ref C)

toggle
connector
(Ref T)

flat
trimmer
(Ref F)

channel
connector
(Ref U)

Figure 1   ACCS Components

Figure 2   Alberfeldy Bridge

Figure 3   Model A19 Tees Viaduct Enclosure

C527/0

Figure 4   Second Severn Crossing Approach Road Bridge

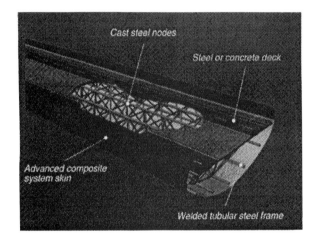

Figure 5   SPACES System

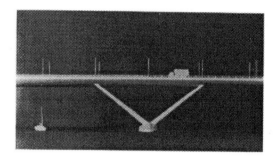

Figure 6   Poole Harbour Bridge Competion Entry

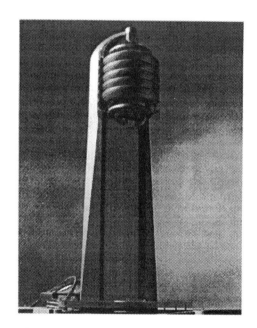

Figure 7   Globorama Tower Model Photo

INNOVATION & WEALTH CREATION

Trent in world markets

D RUFFLES BSc, FIMechE, FRAEng, FRAeS
Director – Engineering and Technology, Rolls-Royce plc, Derby, UK

SYNOPSIS

This paper will review developments in Civil Aero Engines over the last 25 years leading to the Trent engine which powers the A330 and Boeing 777 aircraft. Industrial derivatives for power generation, gas pumping and ship propulsion will also be described together with several of the technological advances which have enabled this family of engines to become world leaders. The paper will conclude with describing possible future developments and their underpinning technologies.

1   HISTORICAL OVERVIEW

The Trent aero engine represents the very best in current engineering technology. The engineering effort in designing the Trent is a very significant undertaking, (Fig.1), and is the latest step in an evolutionary trail that can be traced back over 30 years. The father of the Trent engine unquestionably is the RB211, developed during the late 1960's, 1970's and early 1980's, which pioneered the unique 3 shaft design on which the Trent engine is based, (Fig.2). It is this 3 shaft design which enables short rigid rotor systems and more optimum aerodynamics than 2 shaft engines. This innovation has largely been responsible for the high levels of reliability, performance retention and long life which have characterised our civil engines in the last 25 years. It is because of these qualities that the RB211 has achieved an extraordinary market penetration with well over 3000 engines sold or on order world-wide, (Fig.3). The advantages of the 3 shaft design increase with engine size, by-pass ratio and overall pressure ratio and is a major reason why the Trent 800 is over 1.5 tons lighter than one of our competitors engines, and over 1 ton lighter than the other.

To illustrate the development in aero gas turbines from 1958 to the present day, consider the Avon powered Comet IV, the first jet passenger service across the Atlantic, (Fig.4). Compared with today's aircraft and high by-pass engines there has been a 75% reduction in operating cost per seat mile since the days of the Comet.

Figure 1 The program size

Figure 2 Three shaft layout

This is of course a function of larger and more efficient aircraft, made possible by much more powerful engines which offer significantly better fuel efficiency and reliability. It is interesting to note that the engine improvements alone account for over 65% of this reduction in the form of reduced fuel and maintenance costs, the remaining 35% being attributable to improvements in aircraft design and operation. The dramatic improvements in engine owner-ship costs have been achieved by developing both design and manufacturing technology.

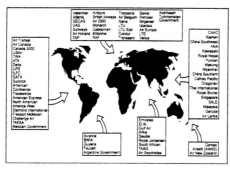

Figure 3 RB211 Market penetration

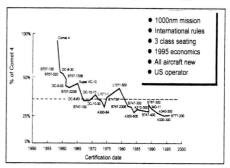

Figure 4 Operating costs per seat mile

2 THE MARKET DEMAND FOR THE TRENT

There has been in excess of a 300% rise in the number of flying passengers in the last 25 years, with over a 450% increase in passenger miles in the same timeframe. It is quite clear from these statistics more people are travelling further than ever before. To satisfy the demand for passenger capacity and range whilst minimising operating costs new aircraft types have been designed with planned derivatives. The Rolls-Royce Trent engine used for the Boeing 777 and Airbus A330 twin engined aircraft has already secured a market share of 33% with more than 300 engines sold and options on a further 150, worth approximately $5.5 billion.

(Fig.5). There are now some 29 aircraft flying powered by the Trent engine, this statistic will be rapidly increasing as more of the engines on order are delivered to our customers.

The airline industry has become a fiercely competitive global business and to survive our airline customers have increasingly demanding requirements of its major suppliers. The aero Trent engine has been designed to meet the demands of thrust, fuel consumption, weight, reliability, emissions, noise, and above all cost of ownership by utilising the very latest in technology.

Figure 5 Boeing 777 and Airbus Industries A330

3   TRENT AERO ENGINE TECHNOLOGY

Design and manufacturing process development in Rolls-Royce has always been a key to our success and the Trent engine embodies these latest technologies, (Fig. 6).

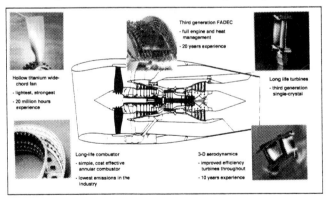

Figure 6 Trent technology

An excellent example of this progress is the wide chord fan blade. (Fig.7). This innovative design results in fewer and stronger fan blades compared with existing solid blades. The wide chord provides the necessary stiffness to dispense with vibration damping snubbers, thus improving aerodynamic efficiency and flow rate. More recently for the Trent engine the honeycomb reinforced hollow titanium wide chord technology has been superseded by a super plastically formed, diffusion bonded (SPF/DB) blade with higher strength, lower weight and lower unit cost. As a result of these technology developments Rolls-Royce has been able to offer greater thrust for a given fan diameter, better fuel efficiency, and better resistance to foreign object damage.

Figure 7 Wide chord fan blade

Computational Fluid Dynamics (CFD) has become a universal tool to the engineer. This technique which allows the modelling of complex fluid flow allows engineers to better understand the design requirements of turbo machinery components. The rapid development of desktop computing power has enabled the engineer to model with ever increasing detail complex 3 dimensional flow processes, visualisation of which has led to new advanced designs for the Trent's gas path blading. These advanced 3D designs which optimise the gas path aerodynamics yield improvements in efficiency and fuel consumption.

Advances in finite element analysis modelling has also allowed for the optimisation of the whole engine structure and mechanical components for mechanical and thermal stressing. Redundant structural material can be avoided in a design, resulting in components that can achieve a higher performance whilst minimising component weight.

Gas turbine combustor design has been advanced to meet the world-wide concern over environmental pollution. Rolls-Royce has reduced emissions from its engines with the Phase 5 combustor in the Trent, (Fig. 8) and can proudly boast the lowest emissions in its class. This world beating technology is acquired through joint industry and university research, again using advanced CFD techniques and in combination with flow visualisation rigs, full rig testing and complete combustor testing in technology demonstration engines. Future combustor technology to further reduce harmful emissions may come in the form of fuel staging, where multiple burning zones tightly control the combustion process.

Materials development is one of our core technological development strategies. The pacing technology of materials dictate the operating temperature and rotational speed of a gas turbine and hence its overall efficiency. Materials also influence the weight of a particular engine which has profound implications to airlines in terms of payload capacity, fuel consumption, and operating range. Optimum material selection is also critical to maximise component life thus minimising the engines through life cost. High temperature technology advances in both materials and manufacturing techniques has allowed a progressive temperature rise of the gas leaving the combustor and entering the high pressure turbine of approximately 500°C over the last 30 years. Casting technology has evolved the early wrought alloys with simple air cooling passages, through directionally solidified cast materials, to today's current technology of complex Nickel alloy single crystal multi-pass cooled blades, with 3D external geometry. It is estimated that a further 200°C increase in turbine entry temperature will be required over the next twenty years to meet the airlines demand for improved performance. This temperature increase is most likely to be accommodated with ceramic thermal barrier coatings which insulate the metal from the hot gas stream and advanced blade cooling made possible by further improvements in casting technology. If a breakthrough in materials such as ceramic matrix composites is made then such materials could be developed to operate without the requirement for parasitic cooling. When embodied into an engine this would improve the cycle efficiency and thus specific fuel consumption. Presently there is no sign of a suitable material becoming available for the hot end turbines and their first application will most probably be for structural components.

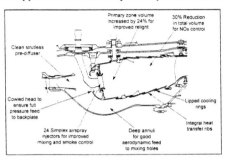

Figure 8 Trent engine phase 5 combustor

Noise from the Trent engine meets all present legislative levels with significant margin. The development of a low noise engine starts at the formation of the initial design where aerodynamic noise generated frequencies are designed out. Also correct selection of rotor and stator blade numbers can cut off or trap sound propagation through the engine. Supplemented with extensive acoustic treatment of the nacelle and engine cowling the technology for noise reduction is far reaching into the design process. Engine noise, however, is only one of the many faceted noise problems surrounding an aircraft. To this end Rolls-Royce are involved with airframers, airline operators, and research establishments in discussing noise reduction possibilities in aircraft design, aircraft operation, and engine design. These collaborative ventures are aimed at jointly further reducing aircraft noise at airports and altitude.

Maintainability of the engine externals and all line replaceable units was a key design goal. To achieve the required maintainability targets and optimise the external layout for the Trent all engine external dressing items such as ancillary systems, electrical harnesses, pipes, and hoses were all electronically 3D modelled using the techniques of digital pre-assembly (DPA). This extremely accurate technique allows optimised routing of external items prior to manufacture, thus saving cost on engine build by eliminating potential non-conformances. In addition, DPA was solely responsible for saving 3 months in the Trent development programme over more traditional techniques of physical mock-up modelling by avoiding the delays that had previously occurred during development build.

Extended Twin Operations (ETOPS) is the aviation industry measure of the flying range between potential airports and therefore routes a twin engined aeroplane is certified to fly. The Trent engine has been designed to satisfy 180 minute ETOPS from entry into service. Traditionally certification for 180 minute ETOPS would only have been achieved with many service hours of running to prove the engines reliability. To achieve the ETOPS certification at entry into service demands an extremely focused design effort to mitigate all past service problems, and design reliability into the engine. Exhaustive testing has been conducted at our no. 57 testbed, (Fig. 9), which subjected the engine to an equivalent of 250,000 flight hours.

Figure 9 Test facility - 57 bed

The 57 test bed was purposely designed to conduct Trent development and ETOPS testing. During the ETOPS tests the engine was mounted by the airframe pylon and all controls were commanded through an aircraft control system. Attention to detail on these tests was extremely high and to fully test the engine even airline service engineers were used to perform "on wing" maintenance during the whole testing process. Following this testing the engine had to undergo actual flight testing. One new production engine and one deliberately aged with testbed flight cycles were mounted on the aircraft and subjected to a 1000 cycle flight test, this testing was aimed at validating the engine/airframe combination, and again included actual airline service engineer maintenance.

4   TRENT DERIVATIVES

4.1 Industrial Trent

Rolls-Royce power generation strategy is focused on complete power plant applications upto 150MW based on its aeroderivative gas turbines supported by its steam turbines and diesel engines. The worlds most powerful and efficient simple cycle aeroderivative gas turbine, the industrial Trent, with a power output of over 50MW at 42% efficiency is the latest addition to the power generation portfolio of gas turbines. It is interesting to note that 50MW is typically sufficient power to satisfy the demand of 50,000 people.

There is a high rate of change in the power generation market. With a growing vertical disintegration of the generation, transmission, and distribution arms of the once electricity companies, the market now recognises the independent operator and not the interdependent operator as was in the past. Full competition status in April 1998 within the U.K power companies will see a growing demand for cost effective power generation capabilities by independent operators. There is therefore a growing market for power systems upto the 150MW class with aeroderived gas turbines engines with their advantages of compact size, high availability, and rapid start up times recognised.

Proven Trent 800 engine components have been used extensively in the industrial Trent with the IP, and HP compressors and turbines common, (Fig.10). Without the need for the fans large bypass flow a two stage low pressure (LP) compressor replaces the aero Trent fan. Minor changes to the LP compressor blades allow for LP generating speeds of either 3000 or

3500 rpm and hence 50 or 60 Hz electrical generating frequencies depending on the power generation application.

Many countries world-wide now have in place legislation which limit the quantity of exhaust emissions such as CO and NO_x from static rotating equipment. The combustor section for the industrial Trent has been designed for dry low emissions (DLE) performance. based on the industrial RB211 DLE technology which has accumulated in excess of 30,000 hours of successful operation in North America and Europe.

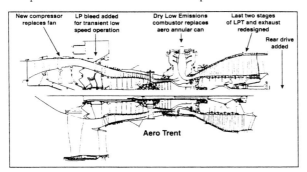

Figure 10 Industrial Trent derivation from aero engine

Eight radial combustors each of series staged design with pre-mixed lean burn have demonstrated NO_x emission levels at 15 volumetric parts per million (vppm) and CO emission levels <1vppm at base load conditions on natural gas operation during development testing. This compares with emissions in the order of 500 vppm of NO_x for the worlds best Phase 5 aero combustor running on kerosene at an equivalent power condition.

The Industrial Trent engine is not only suited to simple cycle operation but also can be packaged with a heat recovery steam generator (HRSG), in a co-generation configuration, utilising waste exhaust heat from the gas turbine which is used to produce steam which can be used in an industrial process plant. The steam production could also be used to power a steam turbine in a combined cycle configuration which is utilised for additional electricity production. Power enhancements of the industrial Trent can be met by intercooling where the engine could be designed to produce 100MW of power, and inlet chilling, whereby the inlet air is chilled to approximately 0°C prior to compression can be used to restore cold day performance of the industrial Trent in hot climates. Complete factory packaging of the industrial Trent into a generating power system is a key design feature, and can be commissioned within 8 weeks at a prepared site. The economic benefits associated with this include reduced project design times, reduced site installation time and capital cost. With flexible power packaging options the customers particular requirements for electrical power and or heat can be met.

The industrial Trent will continue to benefit from the advanced engineering which is embodied into its aero parent, however the evolution does not just stop there. Industrial particular technologies which are being worked include:- emission technologies to <10vppm across wide power ranges and ambient temperatures, advanced corrosion resistant coatings, aerodynamic optimised intake and exhaust systems, complex control system hardware and software to control the DLE combustion process.

4.2 Marine Trent

In the marine market, with collaborative partners Northrop Grumman, we are using Trent and RB211 technology and components, in the development of an advanced, complex cycle gas turbine for the U.S Navy.

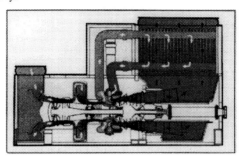

Figure 11 WR-21 Advanced cycle gas turbine

The WR-21 intercooled and recuperated gas turbine, (Fig.11), will provide diesel engine efficiency over a wide speed range, and offer annual fuel savings of 30% over a simple cycle gas turbine, saving $1.5M in fuel costs per ship per year. Just as significant however is the increased capability that a naval vessel will achieve with this improved fuel efficiency, effectively the vessel could spend an additional 8 days 'on station' or add 1000 miles to its operating range, or increase its operational speed by 6 knots. It is also possible to replace the diesel/gas turbine cruise/boost machinery installation typical for some warships with a single WR-21 gas turbine per shaft. This gives a naval vessel significant advantages with reductions to the total ship cost, maintenance, gearbox weight, and installed powerplant volume. With increasing emphasis on enhanced capability at lower cost in the defence sectors, navies around the world understand the significant first cost and through life savings the WR-21 offers. The WR-21 engine system has been designed to fit existing and future naval vessels and is also being considered for the new Anglo/French/Italian 'Horizon' frigate programme.

A marinised Trent engine could also be used for the next generation of sea freighter. This freighter would be able to cross the Atlantic Ocean in three and a half days at speeds in excess of 40 knots, compared with the time of current craft of a week or longer. Innovation of current freighter technology is made possible because of the design of semi-planing monohulls that develop lift, which in itself is not new, but at the same time emerging gas turbine technology in the form of the marine Trent would be capable of delivering the required power to weight ratio, enabling an economically feasible size of vessel to be operated. This type of advanced freighter could be fitted with five marine Trent gas turbines with a total installed power of 250MW (~335,000 hp) powering five waterjet propulsion units. The requirements for this concept are being driven from manufacturers of high value and time sensitive cargoes, to enable goods to reach market place quicker, thus minimising the stock holding requirement and eliminating costs that are associated with the delayed time to market.

5 THE FUTURE FOR THE TRENT ENGINE

Growth versions of the aero Trent to power levels in excess of 100,000lbs of thrust for Boeing 777 long range derivative aircraft are within the capability of the Trent engine. Other new aircraft derivatives such as the Airbus Industries A3XX and A340-500/600 both of which are four engined aircraft designs have their own particular demands for power and range. These new projects can also be powered by Trent derivative engines. The 3 shaft engine design principle enables low risk powerplant sizing, utilising modules from both the Trent 700 and 800 engines with a minimum of engineering redesign, (Fig.12).

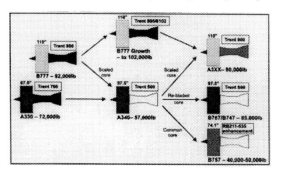

Figure 12 - New Trent derivatives

The forward thinking over 30 years ago when the 3 shaft engine concept was chosen as the architecture for the RB211 has enabled the maximum utilisation of Trent technology for all current aircraft applications. This mixing and matching of Trent modules and technologies will enable us to enhance the -535 and eventually replace the -524. Engine core enhancements of these RB211 engines with Trent technology would see improvements in power, efficiency, and life. The Trent engine design will be further evolved so ensuring that our customers needs of improved performance at a lower cost can be constantly met. Rolls-Royce is dedicated to technology strategies which will embody the best of new and validated technologies at minimum risk and within the necessary cost parameters.

6 CONCLUSIONS

We expect air traffic growth to rise at approximately 5% per annum, which will double revenue passenger miles flown during the next 20 years. However because infrastructure growth is limited this rise in capacity will mostly be met by larger aircraft and some increase in flight frequency. The power demands for these future aircraft can be met by the Trent and is therefore a very important engine to Rolls-Royce. It is generating technology that will be used in our whole family of engines from 2000lbs to over 100,000lbs of thrust and in parallel is also being applied to world beating industrial and marine derivatives.

on-contacting mechanical seals

A PARMAR BTech, PhD, MSTLE
ector of Non-Contacting Seal Technology, John Crane UK Limited, Slough, UK

Traditional mechanical seals on centrifugal pumps are one of the sources of pollution in Chemical, Pharmaceutical and Petrochemical process plants, since they are not able to offer zero emission control of process fluids.

Driven by the need to eliminate these emissions from process plants and to meet increasing demands for seal reliability, John Crane has developed an innovative seal technology for pumps. These new seals utilise a breakthrough gas lubricated, spiral groove non-contacting technology.

The benefits to the plant operator is total process containment ensuring compliance with government and industry standards whilst offering a more cost effective approach than other technologies.

INTRODUCTION

Pumps today are forced to operate under exceedingly harsh conditions. They must pump fluids that are toxic, corrosive, abrasive, or even explosive; and they have to pump them faster than ever before. They must do this while guarding against polluting the atmosphere. Emission reduction has emerged in recent years as a critical issue for seals, for safety and environmental reasons. Conventional mechanical seals cannot always stand up to the challenge.

Consequently, the seal industry responded and a new generation of seals emerged. These new seals utilise a breakthrough gas lubricated non-contacting technology. Originally developed for gas compressors, the concept has been successfully applied to conventional pumps for all types of services.

The innovative concept uses spiral groove non-contacting technology and inert gas (e.g. plant nitrogen) to pressurise a double-seal cartridge. The barrier gas is injected into the seal chamber at 2 bar above the product pressure. At the inboard seal, the patented spiral groove technology pumps a minute amount of gas across the interface into the product, thus providing a barrier against any escape of the product fluid. The outboard seal uses the same non-contacting technology maintaining a controlled leakage of inert gas to the atmosphere.

The benefits of this non-contacting, dry running technology to the plant operator is not only total containment ensuring compliance with government and industry standards, but also offers a much more cost effective approach than other technologies. Considering the additional savings from this energy-efficient and maintenance-free system, the gas lubricated non-contacting seals have the lowest total life cycle costs of the sealing methods available.

Conceptualised and developed by John Crane International, the technology has experienced explosive growth from 1993. The Type 2800 series of non-contacting pump seals have now evolved to cover all pump standards including small bore ANSI and DIN pumps. The design, development and application of world's first gas lubricated non-contacting seals for pumps is discussed in this paper.

NON-CONTACTING TECHNOLOGY

Marked advantages of non-contacting seal technology are rooted in the development of spiral groove technology. Used with a gas barrier system it forms a powerful zero emission seal.

The gas lubricated seal is composed of a stationary ring, typically carbon, which is spring loaded against a mating ring, typically tungsten or silicon carbide, that is mounted on the rotating shaft. Sealing of the gas is achieved at the interface of the rotating and stationary rings in a unique way.

The sealing surfaces are lapped to a high degree of flatness, and one of the rings has a series of spiral grooves machined into the lapped face, figure 1. The design and arrangement of these grooves are such that, with rotation, the gas is pumped into the sealing interface by the grooves viscous shear action. The sealing dam provides a restriction to the pumped flow that results in a pressure build-up between the faces. The magnitude of this hydrodynamic force is designed to balance the seal's applied load at a face separation of a few micrometers.

Figure 2 shows an example of a theoretical prediction of how pressure is built up at the end of the spiral grooves. The results are for a 43mm shaft diameter seal operating at 3000 rev/mm and zero pressure differential.

Under static conditions, hydrostatic forces along with the spring load act to close the faces. Seal balance and the design of the spiral grooves prevent damage to the faces at start-up and shut-down, prior to full separation at normal running conditions.

DEVELOPMENT OF THE TYPE 2800 SERIES OF PUMP SEALS

Application of the spiral groove technology to a dual seal arrangement, with gas as a barrier fluid represents a major paradigm shift in pumping technology. Figure 3 shows a dual cartridge seal that uses gas as the barrier fluid. This is injected into the seal chamber at 2 bar above the product pressure.

A relatively small amount of barrier gas creates the non-contacting feature of this technology. The inboard seal pumps a small amount of gas into the product and the outboard seal maintains a controlled leakage of nitrogen to the atmosphere.

The barrier system is relatively simple, and consists of two parts - a compatible barrier gas source (eg plant nitrogen) capable of providing gas at a pressure of 2 to 3 bar greater than the pump product pressure and a locally mounted gas control panel.

Using the John Crane proprietary advanced computer design tool, CSTEDY[SM] (an integrated package of solids FEA and CFD) the seal geometry and spiral groove features were optimised to meet the performance requirements and to engineer the fit in the limited space available in the pump seal cavity (Ref 1,2).

GAS-LUBRICATED SEAL PERFORMANCE

Extensive laboratory testing was conducted to evaluate the seal arrangements for predictive service life and operational economic values, and to verifi' the performance behaviour predicted by the CSTEDY[SM] design model. Figure 4 shows how the outboard seal leakage increases with increase in pressure.The leakage behaviour is accurately predicted by the CSTEDY[SM] model.

All dual pressurised seals should be capable of operating for a period in the event of loss of barrier pressure, without a reduction in their containment potential. Part of the test programme included simulating the loss of barrier pressure, with the product set at 14 bar. Over a period of two hours of test no reverse transfer of product was detected. This failure condition was also modelled using CSTEDY[SM]. Figure 5 shows the prediction of inboard seal operating in a contacting mode. The contact load spread over the inboard sealing dam contains the product, whilst the outboard seal maintains its non-contacting operation.

GAS-LUBRICATED SEAL BENEFITS AND ECONOMICS

The dual seal cartridge with its non-contacting seal faces offers significant savings in energy. Power consumption can be reduced by a factor of up to 50 compared to a double liquid seal, depending on the operating conditions and the size of the seal. This reduction in power results in reduced operating costs.

Additionally, because of the drastically reduced heat generation, cooling systems can be eliminated and seals can be operated close to the boiling point of the liquid sealed. It follows that the product contamination and clean-up cost associated with liquid support systems are also eliminated. Furthermore, since the seal operates independent of the process liquid being sealed, the seal will not be damaged if the pump is cavitating or operating off its best performance point.

Seal life is not restricted by wear. MTBPM is dramatically increased, with one major chemical plant documenting a 2000 percent increase. Since the seal faces do not contact each other, there is no seal face wear that introduces carbon particles into the product. For certain critical processes, product purity is of great importance - feature which is not guaranteed in a traditional liquid barrier double seal.

Plant operators can achieve similar results by replacing existing pumps with magnetic drive pumps - another zero emission option. For various reasons, including overall investment and capabilities, magnetic drive pumps are inferior to gas lubricated non-contacting seals.

Operating costs of a magnetic drive pump, which may include the cost of electricity and the annualised cost of repair can be 10 to 60 percent higher than for a gas seal. A precise comparison of cost between these technologies depends on the specific application, but it is possible to give general ranges. Table 1 shows the cost for a magnetic drive pump and a gas lubricated non-contacting seal. Various inputs were used, including data gathered during a three year study by a major chemical company. A double liquid lubricated mechanical seal, with an assigned value of 100, formed the basis for comparison.

Table 1 Sealed v Sealless Cost Comparison
(conventional liquid lubricated seal pump = 100)

	Sealless Pump Magnetic Drive	Gas Lubricated Non-Contacting Seal
Installation Costs	80-175	90-110
Operating Costs	110-150	40-90
Cost of Single Repair	175-200	105-150
Mean Time Between Repair	100-250	200-350
Total Life Costs	**125-130**	**65-90**

SUMMARY

The gas lubricated non-contacting Type 2800 series is a unique zero emission product offering customers the following benefits
- Compliance with legislation
- Significant energy savings
- Cost effective solution allowing retrofitting to existing pumps
- Simplifies plant operation

- Increases pump reliability

With over 25 million successful hours of operating experience since its introduction in 1993 , in varied applications that include abrasive, flashing, flammable and toxic fluids, John Crane gas lubricated non-contacting seals represent a more reliable and economical choice for pump users.

REFERENCES

1. Parmar, A., Goldswain I.M., "The Use of Advanced Analysis Techniques In The Design of High Performance Gas Seals", Presented at ACHEMA (1994).
2. O'Brien, A., Parmar, A., Martin, P., "The Application of Zero Emission Dry Gas Sealing Technology to Small Bore Rotating Equipment Standards", BPMA Conference (1995).

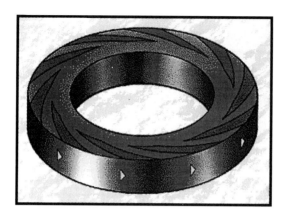

Figure 1

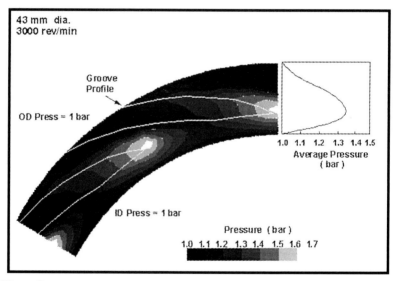

Figure 2

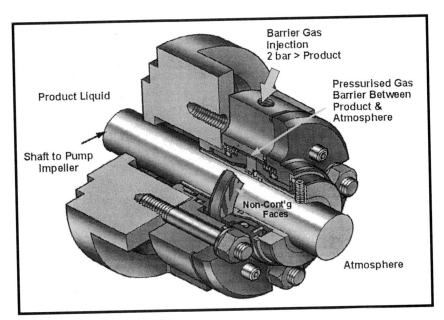

Figure 3

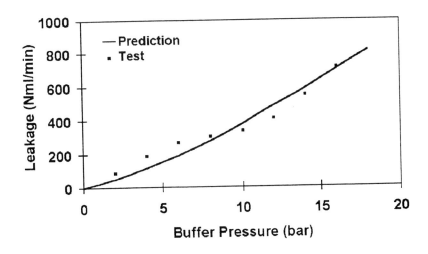

Figure 4

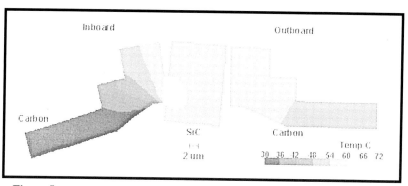

Figure 5

Authors' Index